Physiology of Parasites

TERTIARY LEVEL BIOLOGY

A series covering selected areas of biology at advanced undergraduate level. While designed specifically for course options at this level within Universities and Polytechnics, the series will be of great value to specialists and research workers in other fields who require a knowledge of the essentials of a subject

Titles in the series:

Experimentation in Biology	Ridgman
Methods in Experimental Biology	Ralph
Visceral Muscle	Huddart and Hunt
Biological Membranes	Harrison and Lunt
Comparative Immunobiology	Manning and Turner
Water and Plants	Meidner and Sheriff
Biology of Nematodes	Croll and Matthews
An Introduction to Biological Rhythms	Saunders
Biology of Ageing	Lamb
Biology of Reproduction	Hogarth
An Introduction to Marine Science	Meadows and Campbell
Biology of Fresh Waters	Maitland
An Introduction to Developmental Biology	Ede
Physiology of Parasites	Chappell

TERTIARY LEVEL BIOLOGY

Physiology of Parasites

Leslie H. Chappell, B.Sc., Ph.D

Lecturer in Zoology
University of Aberdeen

A HALSTED PRESS BOOK

John Wiley and Sons

New York—Toronto

Blackie & Son Limited
Bishopbriggs
Glasgow G64 2NZ

Furnival House
14–18 High Holborn
London WC1V 6BX

Published in the U.S.A. and Canada by
Halsted Press
a Division of John Wiley and Sons Inc.
New York

Library of Congress Cataloging in Publication Data

Chappell, Leslie H.
Physiology of parasites.

(Tertiary level biology)
"A Halsted Press book."
Bibliography: p.
Includes index.
1. Parasites—Physiology. I. Title.
[DNLM: 1. Parasites—Physiology. QX4.3 C467p]
QL757.C46 1980 591.5′24 79-20237
ISBN 0-470-26858-1

Filmset and printed in Great Britain by
Thomson Litho Ltd, East Kilbride, Scotland

Preface

THIS BOOK HAS BEEN DEVELOPED FROM A SHORT LECTURE COURSE GIVEN to advanced undergraduate students as part of a general introduction to the subject of parasitology for zoologists.

The book is written for the undergraduate who has no previous experience of parasitology and little background in either biochemistry or physiology. It is not a long book, and students will have to consult some of the more detailed textbooks in parasitology and physiology to gain a full understanding of the topics considered here. My objective in writing this book is to introduce the breadth of parasite physiology, leaving the reader to obtain a depth of knowledge by his own library research.

Each chapter covers a single topic or related topics in physiological parasitology, and the variable length of the chapters reflects the amount of research interest that has been generated over the last few decades. It is to be hoped that by use of this book students will develop an interest in some of the more neglected areas and be stimulated to make good some of the more glaring deficiencies in our current knowledge.

I should like to acknowledge with gratitude the assistance of my colleagues Dr J. Barrett, Dr R. A. Klein, Dr A. W. Pike and Dr R. A. Wilson for reading various chapters, and for their comments. Sincere thanks are due to Bob Duthie for his excellent line drawings and to Alison Wood for typing part of the manuscript. Special thanks go to Eileen for her typing, her encouragement and her patience.

L.H.C.

Contents

Chapter 1.   **INTRODUCTION**   1

What is a parasite?
Historical perspective.
Importance of parasitology to man.

Chapter 2.   **FEEDING AND NUTRITIONAL**   5
PHYSIOLOGY

Introduction.
In vitro culture of parasites.
The alimentary canal of helminths:
     Monogenea.
     Digenea.
     Nematoda.
The role of external surfaces of parasites in their nutrition:
     The morphology of the parasite surface.
Transtegumentary absorption of nutrients:
     Mechanisms of solute entry.
     Transport of molecules into parasites.
     Bidirectional fluxes of nutrients.
Surface enzymes in parasites:
     Intrinsic surface enzymes.
     Extrinsic surface enzymes.
     Inhibition of host enzymes by parasites.
Summary.

Chapter 3.   **CARBOHYDRATE METABOLISM**   42
AND ENERGY PRODUCTION

Stored carbohydrates.
Glycolysis.
End-products of carbohydrate catabolism.
Glycolytic enzymes of parasites.
Carbon dioxide fixation.
Regulation of carbohydrate catabolism.
Tricarboxylic acid cycle.
Role of oxygen in parasite energy metabolism.

Pasteur and Crabtree effects.
Electron transport and terminal oxidations.
Respiratory pigments in parasites.
Pentose phosphate pathway.
Glyoxylate pathway.
Summary.

Chapter 4.   PROTEINS, LIPIDS AND   63
              NUCLEIC ACIDS
Introduction.
Proteins:
    Protein composition.
    Amino acid content.
    Protein synthesis.
    Protein and amino acid metabolism.
    Respiratory proteins.
Lipids:
    Lipid composition.
    Lipid biosynthesis.
    Lipid catabolism.
Nucleic acids:
    Nucleic acid composition.
    Nucleic acid synthesis.
    Nucleic acid catabolism.
    Extranuclear DNA of trypanosomes.
Summary.

Chapter 5.   EXCRETORY SYSTEMS, NITROGEN   82
              EXCRETION, WATER AND IONIC
              REGULATION
Introduction.
Contractile vacuoles in Protozoa.
Protonephridial systems in Platyhelminthes.
End-products of nitrogen metabolism.
Ionic regulation and water balance.
Free amino acids and osmoregulation.
Summary.

Chapter 6.   REPRODUCTION   94
Introduction.
Asexual reproduction.
Sexual reproduction:
    Protozoa.
    Monogenea.
    Digenea.
    Cestoda.
    Acanthocephala.
    Nematoda.
Synchronization of parasite and host reproduction.
Summary.

Chapter 7. PARASITE TRANSMISSION   110
              Introduction.
              Mechanisms for locating the host:
                  Monogenea.
                  Digenea.
                  Cestoda.
              Mechanisms for penetrating the host:
                  Digenea.
                  Nematoda.
              Circadian rhythms and transmission.
              Summary.

Chapter 8. ESTABLISHMENT AND GROWTH   129
            OF PARASITES
              Introduction.
                  Hatching mechanisms:
                  Protozoa.
                  Digenea.
                  Cestoda.
                  Acanthocephala.
                  Nematoda.
              Biochemical aspects of establishment:
                  Role of bile salts.
                  Metabolic changes.
              Migration and site selection:
                  Patterns of migration.
                  Site selection and recognition.
                  Invasion of host tissues.
              Factors inhibiting parasite growth and establishment:
                  Crowding effect.
                  Interspecific interactions.
              Labile growth patterns.
              Moulting in nematodes.
              Summary.

Chapter 9. NERVOUS   SYSTEMS,   SENSE   ORGANS   160
            AND   BEHAVIOURAL   COORDINATION
              Introduction.
              Morphology of the parasite nervous system.
              Sense organs:
                  Monogenea.
                  Digenea.
                  Cestoda.
                  Acanthocephala.
                  Nematoda.
              Nervous transmission, neuromuscular junctions and neuro-
              secretion.
              Behavioural coordination.
              Summary.

Chapter 10. HOST-PARASITE INTERACTIONS   175
        Introduction.
        General principles of cellular and immunological defence
        systems.
        Immunity to Protozoa:
            Blood-dwelling Protozoa
                Malaria.
                Babesia.
                Trypanosoma.
            Tissue-dwelling Protozoa.
                Leishmania.
                Coccidia.
                Amoebae.
        Immunity to helminths:
            Digenea.
            Cestoda.
            Nematoda.
        Immunisation against parasitic diseases.
        Pathogenesis of parasitic infections:
            Immunopathology
            General pathology:
                Mechanical injury.
                Toxic effects.
                Effects on host cell growth.
                Effects on host reproductive systems.
                Metabolic effects.
        Summary.

FURTHER READING   205

GLOSSARY   212

APPENDIX   An outline classification.   215

INDEX   219

CHAPTER ONE

INTRODUCTION

What is a parasite?

THERE ARE MANY DEFINITIONS IN THE LITERATURE THAT DESCRIBE THE intimate associations between animals of different species. Terms like parasitism, commensalism, symbiosis, mutualism, phoresy (and many others) all possess subtle nuances of meaning. There are often arbitrary distinctions between one type of association and another, and it is sometimes difficult to decide on the most appropriate terminology for a particular relationship.

In this book a simplified approach to the problem has been adopted. The term *parasite* is used to refer to those protozoans, platyhelminths, acanthocephalans and nematodes that inhabit, at some time during their life cycle, the body of another larger animal—the host. An outline classification of parasites is given at the back of the book to assist the student with the general comprehension of the text. It should not, however, be regarded as a definitive scheme of classification, and any of the more authoritative texts should be consulted for more detailed information. In general plant parasites have not been included in this book, which examines the relationships between animal parasites and their hosts at the physiological level.

The student should appreciate that the inclusion of the above groups as parasites and the exclusion of the viruses, bacteria, fungi, several minor invertebrate phyla and the arthropods is an expedient rather than an ideal. All these organisms may inhabit tissues of other organisms, and should rightfully be considered together under the term *symbiosis*, which refers simply to the act of living together. The physiological problems of these symbionts are common to all the groups that have adopted this way of life. The limited breadth of the definition of a parasite, as used in this book, is conventional, but for the future we should perhaps consider widening our horizons.

Historical perspective

Physiology, the study of living biological systems, the way they function and the way they adjust to environmental changes, is an ancient branch of the biosciences. By comparison with mammalian physiology, parasite physiology is still in its embryonic stage. During the last three to four decades there has been a gradual increase in the interest shown in this field of research, and, although our knowledge is far from complete, much has been accomplished, and the immediate future promises many exciting developments in the field.

The studies of D. F. Weinland in Germany in the early 1900s are regarded by many as the foundation stone of modern parasite physiology. His investigations of glycogen utilisation, volatile fatty acid production and fermentation in the nematode *Ascaris* provided considerable food for thought for parasitologists over the next two or three decades. During this period, physiologists focussed their attention on a small number of parasitic animals, including *Ascaris*, *Parascaris*, *Fasciola* and *Moniezia* among the metazoans, and termite flagellates, trypanosomatids, trichomonads, rumen ciliates and malaria parasites among the Protozoa.

The prime influence in the development of modern parasite physiology was the Second World War, which exposed huge numbers of individuals to the rigours of life in the tropics, inevitably including the endemic tropical parasites of man. Through the need to develop antiparasite drugs, a knowledge of the biochemistry and physiology of pathogenic parasites became an essential feature of the war effort. Since that time, the subject has become firmly established academically and as an important applied discipline; modern research now utilises both pathogenic and model parasites, and employs the armoury of techniques developed in biochemistry, molecular biology and immunology. There are, nevertheless, areas of parasite physiology that suffer from a notable lack of attention, e.g. neurophysiology, reproductive physiology and excretory physiology. Furthermore, parasitologists have disregarded many of the minor groups of parasites that exist, and have concentrated on those of immediate economic or medical importance. For the interested student, there is much to catch the imagination in the field of parasite physiology and much to be done. The recent contributions of T. Von Brand, E. Bueding, D. Fairbairn, J. F. Mueller, the late C. P. Read, J. D. Smyth and W. Trager, for example, are enormous and should serve as an encouragement to all.

The importance of parasitology to man

Why study parasitology? What is important about increasing our knowledge of parasite physiology?

Parasites inhabit man, his livestock and his crops and they reap an untold harvest of damage. It is estimated that today there are 1000 million sufferers from parasitic diseases. Many of those now infected will die prematurely, countless others will suffer from chronic, painful, debilitating or disfiguring diseases. Parasitic infection of cattle and crops reduces the food resources available for many millions of inhabitants of the Third World. The overall problem is enormous; it demands massive financial backing for research programmes into the control of parasitic diseases and the alleviation of the suffering they cause. First, however, it is necessary for us to increase our fundamental knowledge of the biology of the organisms involved.

Malaria affects 340 million people; in Africa alone, one million children die each year from this parasitic disease of the blood system. Schistosomiasis claims over 200 million sufferers, lymphatic filariasis 250 million, onchocerciasis 20 million. African trypanosomiasis is currently being diagnosed at a rate of 10,000 new cases per annum and a further 35 million people are at risk. New World trypanosomiasis (Chagas' disease) affects 10 million people in Central and South America. Amoebae, hookworms, pinworms, scabies and many other parasites take their toll of human misery. Expensive or ineffective drugs and the development of drug resistance by many parasites exacerbate this massive problem.

From a veterinary standpoint, parasitic infections of livestock dramatically reduce the production of animal protein as a source of food, over much of the globe. African trypanosomiasis of cattle, tick-borne cattle diseases (theileriasis and babesiasis) and the gastrointestinal helminths (fascioliasis, taeniasis and cysticercosis) are major tropical problems, while in the West, nematodes, liver flukes and protozoan parasites (coccidiosis and babesiasis) commonly reduce the production and quality of livestock and cause huge financial losses. On the agricultural front, large scale destruction of crop plants is brought about by infections with plant parasitic nematodes and insect pests.

The essential areas of research are chemotherapy, immunology and *in vitro* culture of parasites, so that their biology may more conveniently be studied in the laboratory. These areas depend upon a fundamental base of physiological knowledge. The World Health Organisation has recently launched a special programme of research and training in tropical parasitology, with these areas of research as stated priorities. The special programme is a long-term, multidisciplinary study of malaria, schisto-somiasis, filariasis, trypanosomiasis and leishmaniasis. One aim is to encourage more parasitologists to investigate the biology of these pathogenic parasites. At present, only three out of every twenty working parasitologists are concerned with parasites of medical, veterinary or agricultural importance, and many universities teach only modest courses in parasitology, even to medical students. Some universities teach no parasitology at all.

Considerable advances in our basic knowledge of parasite physiology have been accomplished over the past three decades, yet we seem more frequently to cite what we do not know than what we know. Today's student of parasitology is in a position to rectify this situation tomorrow, and thereby to make a contribution to the improvement of the quality of life of mankind.

CHAPTER TWO

FEEDING AND NUTRITIONAL PHYSIOLOGY

Introduction

THE ECOLOGIST, CHARLES ELTON, ONCE REMARKED THAT PARASITES LIVE off their capital whereas predators live off their interest—thus suggesting that parasites maintain themselves at the expense of their hosts. In reality, there are very few examples where this conclusion is borne out by experimental observation. Parasites, in general, do not eat themselves out of a home.

The nutritional physiology of parasites is a highly complex subject. Parasites inhabit a wide variety of tissues in both invertebrate and vertebrate animals, including the alimentary canal, the blood, the nervous system, various body cavities and organs, such as the liver and the eyes. Many parasites have exceedingly complicated life cycles, involving up to three different hosts, with two or more free-living stages concerned with transmission, i.e. dispersion and infection of the next host. The transmission stages are often non-feeding phases of the life cycle (e.g. miracidia and cercariae) and there are also non-feeding parasitic larvae that are quiescent, resting stages (e.g. metacercariae). With this wide diversity of form and habitat, it becomes increasingly difficult to make general statements concerning the nutrition of parasites.

Most of the information on the nutritional physiology of parasites derives from studies on the helminth parasites that inhabit either the gut or the blood of vertebrates, and from the blood-dwelling protozoan parasites of vertebrates. These parasites inhabit a microenvironment rich in dissolved organic molecules. Rather less is known about the nutrition of parasites that live within the tissues of their host.

An important concept in the nutritional physiology of parasites, and one that we shall explore in this chapter, is the *host-parasite interface* or *interfacial space*. This is the region of contact between the parasite and its host, and is the site at which the molecular transfer of nutrients occurs. In nutritional terms the interface involves both the external surfaces and,

5

if present, the alimentary canal of the parasite: in the majority of the parasitic nematodes the alimentary canal alone forms the nutritional interface.

In vitro culture of parasites

The culture of a parasite *in vitro* requires the parasite to be grown and maintained outside the host, in conditions that emulate its normal surroundings and support its continued development. The culture of parasites serves two main purposes: to provide parasite material for study in the absence of the host; and as a means of investigating the nutritional requirements of a parasite under completely controlled conditions. In fact, only a small number of parasites have been grown in chemically-defined conditions, containing known quantities of each nutrient. More usually, an undefined addition—such as whole serum, liver extract or yeast extract—is required for successful maintenance of the culture. There is an extensive literature on the *in vitro* culture of protozoan parasites, particularly the haemoflagellates, but rather less is known about the culture of helminth parasites.

Much of the data that we shall discuss in this chapter has been obtained from parasites that are maintained in culture, normally for short periods of time. It must be stressed that a cautious interpretation of these data is essential, since we cannot assume that events measured *in vitro* will accurately reflect the processes that take place *in vivo*. The culture of parasites *in vitro* is a powerful tool for parasitologists but, to be of any lasting service, it requires wisdom in its use.

The alimentary canal of helminth parasites

The Monogenea, Digenea and Nematoda possess an alimentary canal, which varies widely from group to group in both form and function. We shall briefly consider the morphology of the gut in these parasites, and at the same time discuss the importance of this system to the nutrition of parasites. There is no gut in the Protozoa, Cestoda or Acanthocephala.

Monogenea

The gut in the Monogenea conforms to the basic pattern of a mouth, pharynx, oesophagus and either a simple or a highly-branched blind-ending caecal system or intestine (figure 2.1). The mouth is anterior,

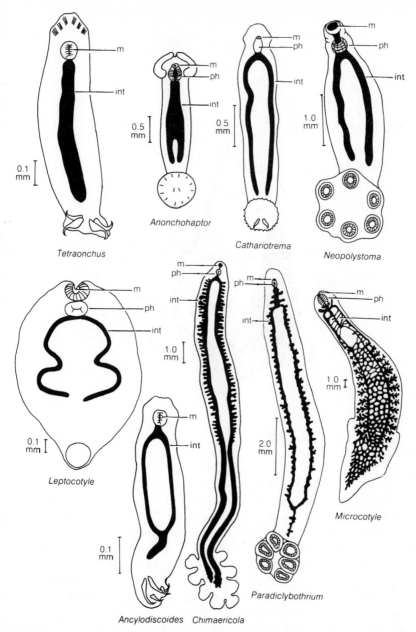

Figure 2.1 The range of form of the monogenean alimentary canal. Based on original drawings by Bychowsky (1957).
m—mouth; ph—pharynx; int—intestinal caeca.

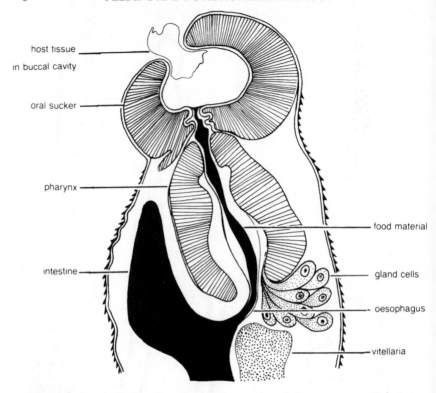

host tissue
in buccal cavity

oral sucker

pharynx

food material

intestine

gland cells

oesophagus

vitellaria

Figure 2.2 Longitudinal section through the anterior of the monogenean *Polystoma integerrimum*, showing the upper alimentary tract and associated glands.
Redrawn from an original photograph by Halton, D. W. and Jennings, J. B. (1965) *Biological Bulletin*, **129**, 257–272.

either at the extreme margin (as in *Microcotyle* and *Polystoma*), or slightly behind the anterior margin on the ventral surface—the latter representing the more usual arrangement. The mouth, which may be simple or elaborated into labia, has associated with it an oral sucker. The buccal cavity leads into the pharynx, a powerful muscular structure composed of three distinct groups of muscle fibres. Glandular structures may be associated with the pharynx (figure 2.2); *Polystoma integerrimum* produces a proteinaceous secretion from its pharyngeal glands, but this species may be unusual among the Polyopisthocotylea in the possession of pharyngeal glands, since other members of the group lack such structures. Within the Monopisthocotylea, some species possess pharyngeal glands while in others the pharynx is wholly muscular. The glandular pharynx of *Entobdella hippoglossi*, for example, reflects the extent to which the pharynx itself is used as an organ of food acquisition by being protruded through the mouth and appended to the surface of the host. In *E. hippoglossi* and *E. soleae*, the secretions from the pharyngeal glands are proteolytic.

The pharynx leads to a short oesophagus, which in some groups (Monocotylidae) is furnished with openings from glands whose function is assumed to be digestive, while in other groups (Microcotylidea and Hexabothriidae) the oesophagus is non-glandular. Little seems to be known about the function of oesophageal glands in the Monogenea.

The intestine extends from the oesophagus to the posterior region of the body and there is normally no anus. The intestine itself varies greatly in form. At its simplest it is a single tube, as in the Tetraonchidae. More usually, however, it is a bifurcated structure. In either case, a substantial degree of secondary branching may occur. According to Bychowsky, the simple form of intestine is characteristic of the smaller monogeneans, while the larger worms tend to possess the more elaborate pattern of intestine, culminating in the extensive anastamoses typical of the large marine monogeneans. Unfortunately, the relationship between the diet and the gross morphology of the monogenean intestine has attracted little attention. Somewhat more information is available, however, on the cellular structure of the gut and its relationship to worm nutrition.

The two orders of the monogenea, the Monopisthocotylea and the Polyopisthocotylea, differ distinctly in their diet. The Monopisthocotylea feed on the epidermal tissues and mucoid secretions of the host, while the Polyopisthocotylea appear to be exclusively blood feeders. This difference is reflected in the cellular construction of the intestinal epithelium, the gastrodermis. (In invertebrates the epithelial lining of the intestine is called the gastrodermis and is analagous to the mucosa of vertebrates.) The tissue and mucus feeders possess a continuous gastrodermis, whereas the sanguinivorous monogeneans are normally characterized by a *discontinuous* or *deciduous* gastrodermis in which the entire epithelium is replaced with every meal ingested—though this is not the case with *Diclidophora*. This condition is also found in some blood feeding Digenea.

There is also a divergence in the digestive processes of the two orders. The Monopisthocotylea digest their food material extracellularly within the lumen of the intestine (or, more rarely, actually outside the body, as in the case of *Entobdella*), while the digestion in the Polyopisthocotylea is both extracellular and intracellular. The digestive

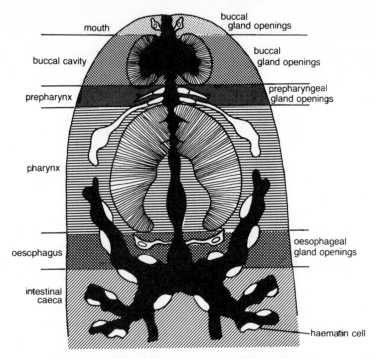

A. THE UPPER ALIMENTARY TRACT WITH ITS ASSOCIATED GLANDS

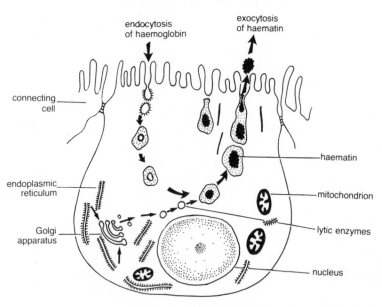

B. THE INTRACELLULAR DIGESTION OF HAEMOGLOBIN IN THE HAEMATIN CELL

processes in the polyopisthocotylean *Diclidophora merlangi*, a common gill parasite of whiting in British waters, have been examined in some detail by Halton and his co-workers (figure 2.3). This fluke feeds on host blood, and digestion of the blood meal follows a biphasic pattern. The extracellular phase commences in the prepharynx, assisted, presumably, by the secretions from at least one of the three glands located in the anterior gut (figure 2.3A). The second phase of digestion takes place intracellularly within the haematin cells of the caeca (figure 2.3B). The caecal gastrodermis is not discontinuous but consists of alternating haematin cells and connecting cells, the latter providing support for the former. Absorption of haemoglobin is accomplished by endocytosis (pinocytosis) across the microvilli of the haematin cells, during which process vesicles are formed. There is a direct numerical relationship between these vesicles and the presence of host blood in the adjacent lumen of the caeca. Once within the haematin cell, haemoglobin is digested by enzymes originating from the granular endoplasmic reticulum, and packaged in the Golgi apparatus. The waste haematin residues are removed from the digestive cells by continuous exocytosis (reverse pinocytosis) rather than the alternative method involving sloughing off of the entire cell layer. A great deal remains to be learned about the digestive processes in *Diclidophora*; morphological studies show that the foregut has great complexity but the physiological significance of this remains to be elucidated.

One word of caution is necessary. The generalisations made in the preceding paragraphs are based on studies of little more than twenty species of monogeneans out of the many thousands that are known. The overall applicability of these statements therefore must await confirmation or refutation by additional investigations in what is a much neglected area of research.

Digenea

The Digenea are separated into two major groups on the basis of the gross structure of the alimentary canal (figure 2.4). The Gasterostomata possess a ventral mouth situated part way along the body and have a

---

Figure 2.3 The anterior gut and the digestion of haemoglobin in the monogenean *Diclidophora merlangi*, from the gills of the marine fish, the whiting. A. redrawn from Halton, D. W. and Morris, G. P. (1975) *International Journal for Parasitology*, **5**, 407–419, B. redrawn from Halton, D. W. (1975) *Parasitology*, **70**, 331–340.

simple sac-like intestine (figure 2.4A), while the gut of the second group, the Prosostomata, closely resembles that of many monogeneans with an anterior mouth, pharynx, oesophagus and paired (bifid) digestive caeca (figure 2.4B-C). The mouth of the prosostome Digenea forms the major organ of feeding—a role partially fulfilled in many Monogenea by the

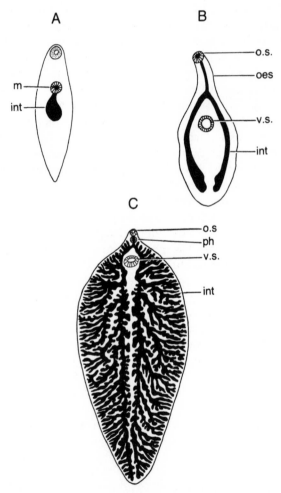

Figure 2.4 The range of form of the digenean alimentary canal. Based on original drawings by Dawes, B. (1968) *The Trematoda*. Cambridge University Press.
int—intestinal caeca; m—mouth; oes—oesophagus; o.s.—oral sucker; ph—pharynx; v.s.—ventral sucker.

pharynx. The pharynx in prosostome digeneans, when present, is a powerful muscular structure that leads to the oesophagus. Some digeneans have glands which open in to the oesophagus but their function is unknown. The basic pattern of paired intestinal caeca (or diverticula) may become modified either by varying degrees of branching or by fusion. There appears to be no obvious relationship between the form of the intestine and the diet of the parasite: e.g. *Haematoloechus medioplexus* and *Haplometra cylindracea* both feed on the blood of their amphibian host and possess a simple bifid intestine (figure 2.4B), while *Fasciola hepatica*, which is probably in part sanguinivorous, has an extensively branched intestinal network (figure 2.4C); the caeca in the schistosomes are anteriorly branched and fuse to form a single structure posterior to the ventral sucker; the Sanguinicolidae, which are blood parasites of poikilothermic vertebrates, are characterized by the possession of an H-shaped intestine.

The caeca in most Digenea are blind-ending, but in some species they open to the outside by way of an anus. In certain echinostomes, for example, the caeca open into the excretory canals and the excretory pore therefore doubles as an anus. Some of the Digenea of fishes (e.g. *Acanthostomum* spp. and *Schistorchis carneus*) have a true anal opening, or even paired anal pores. The functional significance of an anus in the Digenea is obscure, since waste products from feeding and digestion are normally voided via the mouth.

It is rather surprising to find that little attention has been paid to the dietary requirements of digeneans and the best studied species, not unnaturally, are those of economic importance, such as *Fasciola hepatica* and *Schistosoma mansoni*. As with the Monogenea, digeneans ingest either the epithelial tissue of the host or host blood. Sanguinivorous forms include *Azygia* spp., *Haematoloechus medioplexus*, *Haplometra cylindracea* and *Schistosoma* spp. Tissue feeders include *Opisthoglyphe ranae*, *Gorgoderina vitelliloba*, *Pleurogenes* spp. and *Plagiorchis* spp. Some species, such as *Diplodiscus subclavatus* and *Fasciola hepatica*, seem equally able to feed on both blood and other tissues.

The process of feeding in many of the Digenea is assisted by the secretions emanating from glands associated with the ventral sucker, e.g. *Haplometra cylindracea* secretes a non-specific esterase thought to hydrolyse host tissues during feeding; *Opisthoglyphe ranae* produces a mucoid secretion, that may serve to protect the worm from the host digestive enzymes.

The intestinal gastrodermis of digeneans has a range of morphology similar to that seen in the Monogenea. The epithelium is either continuous or discontinuous and, in all species that have been examined, the surface area of the intestine is greatly enlarged by the presence of microvilli. Digestion of food takes place primarily within the lumen of the gut, but this process may be completed intracellularly within the gastrodermal cells following absorption of partially digested material.

Feeding and digestion in Fasciola and Schistosoma Since our knowledge of feeding and digestion in the Digenea depends to a large extent upon studies of these processes in *Fasciola hepatica* and *Schistosoma mansoni*, it is worthwhile considering the nutrition of these two species in some detail.

Schistosomes are exclusively sanguinivorous, and it has long been known that host red blood cells are ingested and haemolysed rapidly within the lumen of the intestinal caeca. The male partner of the permanently paired schistosomes ingests some 30 000 red cells per hour *in vivo*, while the female worm has an appetite approximately ten times greater. This has been demonstrated by the use of radio-chromium labelled red blood cells introduced into infected mice. The schistosome gut is characteristically blackened by the presence of the pigment haematin (produced by haemoglobinolysis) and this waste material is periodically removed by regurgitation; *S. mansoni* completely empties its intestine every three to four hours, the female worm producing four times more haematin than her male partner. Experimental studies on *S. mansoni* in mice indicate that the worm can incorporate amino acids originating from mouse red cell haemoglobin into its own protein, and it has been proposed that peptides, produced by the incomplete lysis of haemoglobin, may be absorbed across the gastrodermis of the worm's gut and incorporated, unaltered, into yolk protein. However, there remains some doubt concerning the precise role of haemoglobin in the nutrition of schistosomes, since only a single proteolytic enzyme has been isolated from the gut of a schistosome and partially purified—this enzyme has a high affinity for native globin and functions optimally at pH 3.9. The products of globin hydrolysis are not amino acids, as might be expected, but small peptides. The female schistosome has a large requirement for proteins and amino acids in her diet as she may produce and liberate up to 300 eggs per day and can do so for the several years of her life span; this rate of egg production costs the female worm ten per cent of her body weight per day. Evidence obtained from the *in vitro* culture of *S. mansoni* supports the view that red cells, or more particularly globin, are necessary for the adequate growth and survival of the worm, i.e. worms will not grow in culture in the absence of red blood cells; the addition of purified globin to the culture increases worm survival. Interest in the nutrition of schistosomes is more than simply academic, since a thorough knowledge of the worm's feeding requirements may eventually allow therapeutic intervention of egg production in infected human beings. This would be of profound importance, as the disease schistosomiasis is primarily associated with the numbers of schistosome eggs that are liberated and that subsequently become trapped within the human body tissues.

Blood feeding in *Fasciola* is less well documented than in *Schistosoma* and is in fact disputed by several authorities. It is probably reasonable to assume that *Fasciola* can and does ingest both epithelial tissues of the bile duct and red blood cells. It has been suggested that the liver fluke browses upon the hyperplastic epithelium produced in the bile duct as a reaction to the presence of the parasite. *In vitro* studies carried out on the feeding of *Fasciola* show that the worm is capable of digesting erythrocytes (red blood cells) and excreting iron as a waste product. Digestion is both extracellular and intracellular, in contrast to the schistosomes where red cells are digested solely in the intestinal lumen. A curious feeding cycle occurs in *Fasciola* during which the gastrodermis is seen to undergo radical morphological alteration. The gastrodermis is highly irregular in appearance, the cells ranging in shape from columnar to squamous—the starved worm is characterized by a columnar gastrodermis in which digestive enzymes (non-specific esterases) are located in the basal region of the cells (figure 2.5). These enzymes gradually accumulate in the apical portions of the gastrodermal cells and, on ingestion of a blood meal, the enzyme-rich cells rupture at their apices and release blebs of enzymes into the lumen of the intestine. These "packets" of enzymes are rapidly mixed with the gut contents and digestion proceeds. The actual presence of food in the gut appears to be the stimulus for the release of digestive enzymes. Partially digested haemoglobin is fully hydrolysed in the gastrodermal cells, and the major waste material, iron, is transported through the parenchyma and removed via the excretory system. Schistosomes solve the problem of unwanted iron from the digestion of haemoglobin rather differently, by regular regurgitation of gut contents; this is usually considered to be a more advanced adaptation to the sanguinivorous habit. In schistosomes the gastrodermis is non-deciduous, whereas in *Fasciola* there is a high rate of turnover of gastrodermal tissue along with the need to excrete

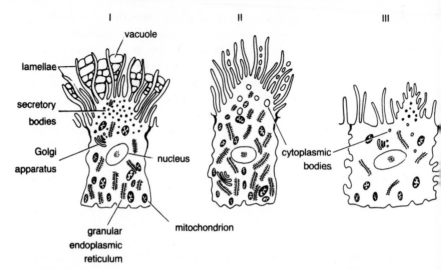

Figure 2.5 Variation in the morphology of gastrodermal cells lining the intestine of the liver fluke, *Fasciola hepatica*. Redrawn in simplified form from Robinson, G. and Threadgold, L. T. (1975) *Experimental Parasitology*, **37**, 20–36.
Group I cells: most abundant type, occurring in clusters of 4 to 7 cells; columnar cells 25–40 µm tall; apices with numerous lamellae separated by vacuoles; granular endoplasmic reticulum abundant; secretory granules increase in number from early to late development.
Group II cells: columnar cells 20–35 µm tall; apices lack secretory bodies but with numerous cytoplasmic bodies containing coiled membrane material; numerous lamellae separated by vacuoles; in late development of this cell type there is a reduction in the number of lamellae.
Group III cells: cuboidal cells, 10–20 µm × 10–20 µm; short lamellae; numerous mitochondria; apical secretory bodies present.
Group I cells are predominantly secretory in function and Group II cells are predominantly absorptive in their function. The sequence of cell types from I to III represents a cycle of functional activity in a single gastrodermal cell.

iron from the body tissues. There is some evidence that *Fasciola* has proteolytic enzymes which hydrolyse globin to free amino acids, the enzymes functioning optimally at a low pH. Clearly we need to extend our knowledge of the gut physiology of these two digeneans in particular and other representatives of the group in general, but the practical problems, though not insuperable, do offer a serious challenge to the researcher.

Nematoda

The alimentary canal of both free-living and parasitic nematodes conforms to a single basic pattern, comprising three distinct regions: the stomodaeum, which includes the mouth, labia, buccal cavity and pharynx; the intestine, which is a straight tube and is the site of digestion and absorption; and the proctodaeum, which includes the rectum and anus. The stomodaeum and proctodaeum are lined by cuticle which is therefore shed with each successive moult. The anterior region of the nematode gut shows the greatest degree of morphological variation and, in general, this reflects the feeding habits of the worm. The oral region of parasitic nematodes is commonly less elaborate than that found in the free-living forms but is complex in groups such as the hookworms. Nematodes feed by the suction produced by the powerful pharyngeal musculature and by "teeth", in those forms that possess them, such as the hookworms (*Ancylostoma*) parasitic in the mammalian gut. The intestine is lined with a microvillous gastrodermis that is both secretory and absorptive.

The animal parasitic nematodes can be divided into four groups according to their feeding habits: feeders on the contents of their host's gut (e.g. *Ascaris*); epithelium feeders (e.g. *Ancylostoma*); tissue feeders involving tissue penetration (e.g. *Trichuris*); and fluid tissue feeders (e.g. *Wuchereria*).

There is virtually no evidence for the transcuticular absorption of nutrients in nematodes and the gut must be regarded as the functional interface. Exceptions to this rule are the entomophilic (insect-parasitic) nematodes, such as *Mermis nigrescens*, parasitic in the haemocoel of locusts, and *Bradynema*, a parasite of dipterans; in the latter species the mouth, anus and gut degenerate and the cuticle develops a microvillous surface, through which nutrients can pass, while the cuticle of the former is permeable to nutrients despite the presence of the alimentary canal.

The role of the external surfaces of parasites in their nutrition

The external surfaces of all endoparasites form a potential nutritional interface between the parasite and its host. In the case of the Cestoda and the Acanthocephala, the complete absence of an alimentary canal renders the body surface of the parasite the major site of nutrient uptake, while the surfaces of all other parasites, with the virtual exception of the Nematoda, may play a significant nutritional role.

The evidence for the nutritional role of parasite surfaces comes from both morphological and physiological studies, particularly investigations of membrane transport of small organic molecules and ions. We shall consider both of these aspects in turn.

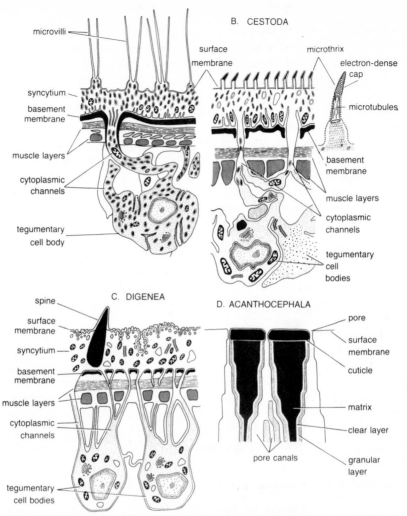

Figure 2.6 The structure of the surfaces of helminth parasites (excluding nematodes) based on ultrastructural studies using the transmission electron microscope.

Platyhelminths are covered by a metabolically active layer and not a cuticle as was formerly thought to be the case; the acanthocephalans and nematodes do possess a cuticular covering, which, in the former group, is richly supplied with pores. The Protozoa are covered simply by a plasma membrane.

A. redrawn from Lyons, K. M. (1970) *Parasitology*, **60**, 39–52.

B. redrawn from Smyth, J. D. (1969) *The Physiology of Cestodes*. Oliver and Boyd.

C. redrawn from Threadgold, L. T. (1963) *Quarterly Journal of Microscopical Science*, **104**, 505–512.

D. redrawn from Crompton, D. W. T. (1970) *An Ecological Approach to Acanthocephalian Physiology*. Cambridge University Press.

The morphology of the parasite surface

Until the advent of the electron microscope, it was generally accepted that platyhelminth parasites were covered by a cuticular surface which, by analogy with the cuticle of other invertebrate groups, was regarded as an impermeable and protective structure. We now know that this is not the case, and instead we find that the surface covering of these worms is a metabolically-active syncytium whose cell bodies are sunk deep into the underlying parenchymatous tissue of the worm. The surface is thus called a tegument, though the terms integument or epidermis appear to be interchangeable. The similarities between the tegumentary structure in the Monogenea, Digenea and Cestoda encourage us to consider them together (figure 2.6).

The platyhelminth tegument is composed of two distinct layers: an outer anucleate syncytium, and an inner nucleated region. The syncytial region is bounded externally by the plasma or surface membrane, and internally by the basement membrane. Outside the surface membrane lies the *glycocalyx* or "fuzzy coat", a polyanionic layer of varying thickness, and rich in carbohydrate—the dimensions of the glycocalyx seem to vary according to the fixation techniques employed. Studies using radioactive tracers suggest that the glycocalyx and the surface membrane may arise in vesicles formed by the Golgi apparatus of the tegumentary cells. The gross surface topography is highly variable in the helminths, not only from group to group but in different regions on the body of an individual worm, e.g. in *Fasciola hepatica*, some regions of the surface show deep invaginations while other areas are quite smooth. The surface of the Digenea is typified by the presence of spines which are embedded into the syncytial region of the tegument and are covered, externally, by the plasma membrane. The density of these spines frequently differs over the surface of the worm and they tend to be least numerous posteriorly. The cestode surface is unusual in the possession of numerous microvillous projections from the surface called *microtriches*; these structures are found in all groups of tapeworms that have been examined, from the most primitive Proteocephalidea to the Cyclophyllidea. In general, the density of the microtriches is lower in the scolex and neck region of the worm than in the more mature regions of the body. The cestode microtriches resemble the brush border found on many cell surfaces (e.g. the mammalian mucosal cell), but they differ in the possession of electron-dense tips, the function of which remains obscure. While the microtriches serve to increase the surface area of the tapeworm (by a factor of 3–6 times) they are not as effective in this regard as are the microvilli on the mammalian mucosal cell and it may be that they have alternative functions, such as attachment and maintenance of position in the host gut. Microtubules run longitudinally within the shaft of the microtriches and these probably have a function in the translocation of nutrients. Rather surprisingly, the surface of monogeneans is also microvillous despite their external position on the host and this may be best explained on phylogenetic grounds rather than in functional terms, since it is unlikely that external parasites would make use of their surfaces for nutrient acquisition to any significant extent.

In all the platyhelminths the syncytial tegumentary layer, frequently referred to as the distal or perinuclear cytoplasm, is richly supplied with mitochondria which have characteristically few cristae, endoplasmic reticulum, various vacuoles and vesicles that contain replacement surface membrane and glycocalyx. These vesicles have been studied in detail in *Schistosoma mansoni*, a digenean with a complex surface structure whereby the membranes of the surface proliferate during the development from larva to adult in the mammalian host. The plasma membrane of the schistosome cercaria, the invasive larva, is trilaminate but it becomes heptalaminate and deeply folded within three hours of penetration of the host. There is also thought to be a continual process of surface membrane turnover in these worms.

The tegumentary cells of platyhelminths are located within the parenchymal tissue below the circular and longitudinal muscle layers. The cytoplasm of these cells is connected to the syncytial layer by an extensive network of tubules. The cellular region of the tegument contains the nuclei, Golgi apparatus, mitochondria and endoplasmic reticulum. It is probable that all platyhelminths have a relatively high turnover rate of surface membrane, with new membranous material assimilated within the nucleated portion of the tegument.

The overall structure of the platyhelminth tegument, as revealed by electron microscopy, is clearly consistent with a nutritional and metabolic function rather than simple protection. As we shall discuss later there is an accumulating body of evidence to show that, in the Cestoda and Digenea and to a lesser extent the Monogenea, the tegument is an active region of nutrient uptake.

The body surface of adult Acanthocephala is rather different from that of the parasitic Platyhelminthes, comprising five distinct layers. Externally lies a thin mucopolysaccharide-containing epicuticle, internal to which is a cuticle, a striped layer, a fibrous layer and an innermost radial layer, residing on the basement membrane. The surface region, or body wall, is syncytial and metabolically active, as indicated by the presence of nuclei, mitochondria, ribosomes and folded plasma membranes, and it is rich in organic material such as glycogen and lipid. An interesting feature of the acanthocephalan surface is the presence of numerous pores that penetrate through the cuticular layers and extend throughout the body wall. Each pore is lined with plasma membrane and is enclosed within a granular layer, itself contained within what appears to be a clear zone. The canals derived from the pores divide considerably as they extend internally serving to provide an enormously increased surface area. It is not known how these pores and their canal system are involved in nutrient acquisition in the Acanthocephala, but their role must be important considering the absence of a gut and the impermeable nature of the cuticular surface. We shall discuss the evidence for membrane transport in the Acanthocephala later.

The great majority of nematodes are covered by a thick, often multiple layered, cuticle which is impermeable to solutes. The cuticle of *Ascaris* comprises three major regions: an outer cortex, which may contain keratin and collagen; a central matrix, which is made up of a number of proteinaceous layers; and an innermost fibrous layer bounded by a basement membrane. The matrix layer of *Nippostrongylus brasiliensis* is fluid-filled and is rich in both enzymes and haemoglobin. There is no evidence to suggest that the nematode cuticle functions as a site for nutrient absorption, except in the entomophilic species.

For most nematodes, this activity must reside exclusively with the gut and its associated structures.

Transtegumentary absorption of nutrients

In discussing the functional morphology of the alimentary canal of parasites (where it is present) it will have become evident that current knowledge is far from complete concerning the digestive and absorptive activities of the gut. We can probably assume that the gut in the Monogenea and Digenea is involved in the digestion and subsequent absorption of host tissues and red blood cells, but studies on the occurrence and activity of alimentary enzymes suggest that the gut may be somewhat limited in its potential, e.g. *Schistosoma mansoni* possesses, as far as we can tell, only a single intestinal proteolytic enzyme, hydrolysing globin to peptide fragments. One can reasonably question, therefore, the precise origin of free amino acids for somatic protein synthesis. We cannot provide an answer to this question at present, but we are in a position to suggest that the transtegumentary absorption of low molecular weight nutrients, known to occur, may be of great significance to schistosomes, and indeed all internal parasites, with the notable exception of most nematodes—obviously it must represent the major route in the Cestoda and Acanthocephala where there is no gut.

Direct evidence for the absorptive capabilities of the external surfaces of parasites has been obtained by studying the membrane transport of radioactively labelled solutes *in vitro*. Surprisingly, as a result of such studies, more is known about tegumentary absorption than that occurring through the intestine of parasites. This is probably due to the excellence of the adult tapeworm as an experimental system, exploited with considerable success by Read and his colleagues during the last fifteen years. There have been many similar transport studies on the parasitic Protozoa, particularly the trypanosomes.

Mechanisms of solute entry

The plasma membrane of all cells serves not only to maintain the integrity of the cell but also to regulate the passage or flux of dissolved or insoluble molecules. The chemical structure of the plasma membrane has been the focus of considerable attention for many years and a number of models have been proposed, none of which has gained complete general acceptance. The major membrane components are lipid (60%) and protein (40%), and there is good evidence that the lipid exists in the form of a bilayer with the hydrophobic ends of the molecules orientated to the inside of the membrane. The relationship of the protein component to the lipid bilayer is in some doubt, but proteins probably extend through the lipid layer.

As we have discussed already all parasites, with the exception of nematodes, are bounded by an external plasma membrane and there is no evidence suggesting that this is unique in its structure in parasitic animals. Therefore we may assume that the surface membranes of parasites provide selective barriers to solutes and they may be studied using techniques and approaches developed for other transporting systems.

Solutes move across membranes by a number of different processes: simple diffusion, facilitated or mediated diffusion, active transport and pinocytosis (endocytosis or exocytosis). These modes of transport are readily distinguishable by their biochemical and kinetic characteristics which can be determined under *in vitro* conditions in the laboratory.

Simple *diffusion*, sometimes termed passive transport, obeys Fick's law which states that the rate of diffusion of a solute is directly related to the concentration difference on either side of the membrane. A solute will always migrate from a region of high concentration to a region where the concentration is lower. When the concentration of the solute is the same on both sides of the membrane the net flux of the solute is zero. Fick's Law can be expressed as follows:

$$\frac{dn}{dt} = DA\frac{dc}{dx}$$

where dn/dt is the number of molecules of the solute (dn) moving across the membrane with a surface area A in the time dt, and where a concentration difference of dc exists over the distance dx. D is the coefficient of diffusion. Movement of a solute by simple diffusion requires no energy expenditure on the part of the cell, is unaffected by other solutes or by metabolic poisons and shows no predilection for a particular stereoisomer.

Facilitated or *mediated diffusion* (figure 2.7) differs from simple diffusion in the involvement of a binding process at some point during transport and hence it exhibits saturation kinetics. Frequently the term "carrier" is used to indicate the unknown structure that binds the solute and transports it across the membrane, releasing it at the appropriate place on the other side. However, since there is little convincing evidence for the existence of these carriers *per se*, it is probably preferable to use alternative terminology. Parasitologists prefer to use the term *site* or *locus* in place of carrier, whereas other workers have employed terms such as *permeases*. In this book we shall use the terms site or system, thereby leaving open the nature of the transport process. Facilitated diffusion, like simple diffusion, involves solute movement down an electrochemical gradient, so that no accumulation of solute on one side of the membrane is possible. Accordingly, no energy is used to transport molecules by facilitated diffusion. The graphical relationship between the concentration of the solute and its rate of transport in facilitated diffusion is non-linear, and as the concentration of solute increases so the rate of uptake approaches an asymptote. Solute transport by facilitated diffusion can be inhibited by a variety of chemically similar molecules which compete or interfere with the binding of the solute at the transport site. The degree of specificity of each transport site may vary widely, as we shall see, being a facet of the array of molecular species that inhibit

2. MEDIATED TRANSPORT

1. SIMPLE DIFFUSION

(FACILITATED DIFFUSION
OR ACTIVE TRANSPORT)

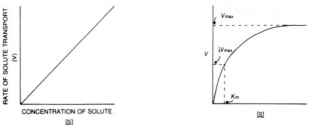

3. LINEWEAVER-BURK DOUBLE
RECIPROCAL PLOT

(BASED ON THE DATA IN GRAPH 2)

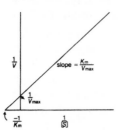

4. COMPETITIVE INHIBITION OF TRANSPORT

(V_{max} REMAINS CONSTANT)

5. NON-COMPETITIVE INHIBITION OF TRANSPORT

(K_m REMAINS CONSTANT)

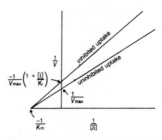

Figure 2.7 The kinetics of membrane transport. In practice, the uninhibited uptake of low molecular weight moieties by intact parasites *in vitro* often fits a curve which is a combination of 1. and 2. above. Mediated transport is more important at low substrate concentrations, but at higher concentrations more substrate enters by simple diffusion.
K_i—inhibitor constant (can be determined experimentally as in 5. or 6.); [I]—concentration of inhibitor; K_m—transport or Michaelis constant; V_{max}—maximum rate of transport ($K_m = \frac{1}{2} V_{max}$).

uptake of a particular solute. Stereospecificity is characteristic of facilitated diffusion. An advantage of this type of transport process is that the rate of solute movement may be much greater than could occur by simple diffusion.

Active transport is kinetically similar to facilitated diffusion, but in the active process solutes are transported *up* an electrochemical gradient and metabolic energy is expended by the cell. Thus active transport permits concentration or accumulation of a solute inside a cell when the initial concentration of the solute inside the cell exceeds that on the outside. Active transport frequently involves the simultaneous movement of sodium ions, and solute transport can often be interrupted by altering sodium transport, e.g. by adding ouabain. Active transport also shows stereospecificity and various degrees of inhibition as in facilitated diffusion, but differs in being sensitive to metabolic poisons.

Pinocytosis, often called endocytosis or exocytosis depending upon the direction of the movement, is a macromolecular transport process whereby large molecules (e.g. proteins) are transported in membrane-bound vesicles. Pinocytosis may show saturation kinetics since it involves the initial binding of the macromolecule to the membrane prior to vesicle formation. Metabolic inhibitors may interefere with transport by this process.

An important consideration, germane to discussion of transport of molecules into parasites, is the use of intact organisms as opposed to the wide use of isolated tissues, e.g. Erhlich ascites tumour cells, bacterial preparations or isolated intestinal loops or rings, for the many transport studies,in non-parasites. As a consequence the whole-animal approach used by parasite physiologists lends itself to problems of interpretation not encountered in the isolated cell systems, particularly with regard to the internal concentrations of naturally occurring molecules, and no account can be taken of concentration differences within the various body tissues of the intact parasite. Compartmentalization of molecules is a problem which has not been overcome to any satisfactory degree. Ideally, transport of solutes across the isolated helminth tegument should be examined, but the diffuse structure of the surface in metazoan parasites precludes this—this criticism, however, does not apply to the protozoan parasites. Nevertheless, a great deal of important information on transtegumentary transport in helminth parasites has been obtained by the use of carefully controlled experimental conditions.

Michaelis-Menten kinetics: In 1913 Michaelis and Menten formulated an equation to account for the rate of an enzymic reaction. The equation, known universally as the Michaelis-Menten equation, can be expressed as follows:

$$V = \frac{V_{max}[S]}{[S] + K_m} \quad \text{or} \quad V = \frac{V_{max}}{\dfrac{K_m}{[S]} + 1}$$

where V is the rate of the reaction, V_{max} is the maximum velocity and K_m the Michaelis constant or concentration of substrate S required to yield half maximum velocity. This equation applies to a saturable enzymic reaction and assumes the reversible formation of a complex between the substrate and the enzyme:

$$E + S \rightleftharpoons ES \rightleftharpoons E + P$$

where E is the enzyme, S the substrate, ES the transition complex and P the product of the reaction. It has been subsequently noted that this equation will apply equally well to the saturable process of membrane transport, such as occurs in facilitated diffusion and active transport:

$$C + S \rightleftharpoons CS \rightleftharpoons C + S$$

where C is the hypothetical membrane carrier, S the substrate transported, and CS the presumed complex formed between the carrier and the substrate, in which the substrate is presumed to be transported across the membrane. There are two important differences concerning the Michaelis-Menten equation when applied to membrane transport rather than to an enzymic reaction: first, there is probably no chemical change of the substrate during its transport, and secondly, unlike enzymes, membrane carriers await identification and isolation and must therefore be regarded as hypothetical entities whose presence is only implied by the occurrence of saturable transporting systems.

While the application of the Michaelis-Menten equation to transport systems provides no insight into the actual nature of the transport process itself, it does provide a very convenient means of comparing the kinetic parameters of different systems, or similar systems in different organisms.

Transport of molecules into parasites

Carbohydrate transport   The uptake of carbohydrates has been investigated in only a small number of parasites including three species of haemoflagellate Protozoa (*Trypanosoma lewisi*, *T. equiperdum* and *T. gambiense*), two species of Digenea (*Fasciola hepatica* and *Schistosoma mansoni*), four species of Cestoda (*Hymenolepis diminuta*, *H. microstoma*, *Taenia crassiceps* and *Calliobothrium verticillatum*), two Acanthocephala (*Polymorphus minutus* and *Moniliformis dubius*) and across the intestinal wall of two species of Nematoda (*Ascaris suum* and *Trichuris vulpis*). The detailed review of transport into parasites by Pappas and Read (1975) is recommended for additional reference.

The cestode *Hymenolepis diminuta* has provided an ideal model system for transport studies since it can be maintained in the laboratory with ease, is not pathogenic to man and, as a system, it yields highly reproducible data. Not surprisingly, considerable information has been amassed on the transport of nutrient molecules into *H. diminuta*, while much less is known about other species of helminth parasite. Even so, most studies on *H. diminuta* have utilized only ten-day-old worms; these worms are immature, since egg production does not commence until the sixteenth day in the rat gut. Little is known about transport in mature worms or in the younger stages, and there have been just two studies on transport in the cysticercoid larvae obtained from the intermediate host (a flour beetle).

In vitro, glucose is actively transported across the tegument of *H. diminuta*. Absorption of glucose is competitively inhibited by some sugars (D-allose, α-methyl glucose, 6-deoxyglucose, 3-O-methylglucose and 1-deoxyglucose) but not by others (fructose, L-fucose, 1-5-anhydro-D-mannitol) (see table 2.1). The uptake system is stereospecific, temperature dependent, with a Q_{10} (for the temperature range 15–40°C) of 2.4, and is inhibited by metabolic poisons (p-mercuribenzoate, iodacetate and 2, 4-dinitrophenol). The worm can accumulate glucose—if incubated in 5 mM glucose for 60 minutes, an internal concentration of 25 mM is attained. It has been proposed that galactose enters *H. diminuta* via the glucose transport system. The cysticercoid larvae of *H. diminuta* possess a comparable transport system for hexoses which is not surprising in view of the structural similarities between the adult surface and the larval cyst wall. The bile duct tapeworm of mice, *H. microstoma*, also actively transports glucose.

In vivo studies on carbohydrate transport by *H. diminuta* have also employed isolated intestinal loops removed from infected rats. Three concurrent entry processes are reported to occur using this technique: these are active transport, simple diffusion, and a third process known as *solvent drag*. The relative contributions of each vary with alterations of pH. It is not possible to compare directly the transport of carbohydrates *in vitro* and *in vivo*, but it should be stressed that, while the use of an *in*

Table 2.1 Selected kinetic data on carbohydrate transport in cestodes

Tapeworm species	Molecule transported	K_t (mM)	V_{max} (μmol g^{-1} h^{-1})	Inhibitors
Hymenolepis diminuta	glucose	0.7–1.6	400–789	phlorizin galactose α-methylglucoside 2-deoxyglucose 3-O-methylglucose 6-deoxyglucose allose
	galactose	5.0	—	phlorizin
	glycerol	0.2–0.4	84–91	1, 2-propanediol α-glycerophosphate
H. microstoma	glucose	2.0	1080	galactose phlorizin phloretin α-methylglucoside 1-deoxyglucose
Taenia crassiceps (larvae)	glucose	0.3	20	α-methylglucoside β-methylglucoside galactose 6-deoxygalactose phlorizin
	galactose	0.8	70	—
Calliobothrium verticillatum	glucose	0.6	610	α-methylglucoside galactose phlorizin

Data from Pappas and Read (1975)

vitro approach is of undoubted value for examining isolated systems, it may tell us little about what takes place *in vivo*, where great chemical complexity prevails.

Hymenolepis diminuta absorbs glycerol by both facilitated diffusion and simple diffusion. The mediated system is inhibited by 1, 2-propanediol, α-glycerophosphate, iodoacetate and arsenite; uptake of glycerol is also affected by the presence or absence of Na$^+$, by temperature variation and by the pH of the medium. Glucose uptake in all cestodes examined is dependent upon the concentration of Na$^+$.

Carbohydrate transport has been investigated in only two species of the Digenea, *Fasciola hepatica* and *Schistosoma mansoni*. Glucose, fructose, mannose, glucosamine and ribose enter *F. hepatica* by mediated transport processes, probably facilitated diffusion, and xylose enters by simple diffusion. Glucose, galactose, 2-deoxyglucose, glucosamine, mannose, ribose and 3-O-methylglucose (a non-metabolizable analogue of glucose) enter *S. mansoni* by simple diffusion and a mediated saturable system. Cestodes and digeneans may differ in their permeability to fructose, *H. diminuta* being impermeable and the Digenea studied permeable, to this hexose.

Moniliformis dubius (an acanthocephalan) has saturable transport systems for several sugars (glucose, 2-deoxyglucose, mannose, N-acetylglucosamine, 3-O-methylglucose, fructose and galactose). Glucose and 2-deoxyglucose share the same site for entry. Glucose entry is not dependent upon the presence of Na^+ and is not inhibited by phlorizin, in contrast to glucose transport in other helminths. *Polymorphus minutus* also absorbs glucose by a system that exhibits saturation kinetics and may possibly be active transport.

Carbohydrate transport has been examined in a small number of parasitic Protozoa and particular attention has been paid to haemoflagellates of the genus *Trypanosoma*. The mammalian bloodstream forms, *T. gambiense* and *T. lewisi*, transport a variety of carbohydrates by saturable systems. *T. gambiense* transports glucose, mannose, glycerol, 2-deoxyglucose and fructose but not the glucose analogue 3-O-methylglucose. Kinetic data suggest that there are two carbohydrate transport systems in *T. gambiense*; the "glucose site" and the "fructose site", the latter exclusively translocating fructose and glucosamine. Fructose transport is inhibited by a number of carbohydrates (glucose, glycerol, glucosamine, mannose and N-acetylglucosamine) which themselves are not transported through this site. This phenomenon is called *non-productive binding*, where molecules bind to, but are not translocated by, a transport system. In *T. gambiense* the glucose site is markedly specific for the hexose carbon atoms in the 1, 3, 4 and 6 positions (demonstrated by measuring inhibitory efficiency as a function of molecular structure) and the "fructose site" requires carbon atom number 5 of the sugar to remain unaltered. *Trypanosoma lewisi* also has more than one carbohydrate transport system; in this organism, fructose and 3-O-methylglucose inhibit glucose transport, in contrast with *T. gambiense*. There is only a single carbohydrate transport system in *T. equiperdum*. There is some evidence that the age or state of development of parasitic organisms may affect the rate of transport of carbohydrates. This has been demonstrated in *Hymenolepis diminuta* and *T. gambiense*.

Amino acid transport   Unlike carbohydrates, amino acids are not usually involved in the energy metabolism of parasites. Their importance lies in their role in protein synthesis, which is commonly high in helminth parasites due to the large numbers of eggs produced. Amino acid transport has been studied extensively in the cestode *H. diminuta* by Read and his co-workers, while the Digenea, haemoflagellate Protozoa, and Acanthocephala have received rather less attention.

Amongst the Protozoa, studies on lysine transport in *Trypanosoma cruzi*, the organism responsible for Chagas disease, reveal some departures in site specificity from those typical of other organisms. Notably, in the culture form of *T. cruzi* (the stage analogous to the insect stage of the parasite in nature), uptake of lysine (a basic amino acid) occurs through three kinetically distinguishable sites. Moreover, other basic amino acids do not share these routes of entry with lysine. Inhibition of lysine uptake is effected by neutral amino acids as follows: at site 1, phenylalanine, tryosine; at site 2, proline, alanine, methionine and cysteine; and at site 3, glycine, valine, leucine and isoleucine. The partially competitive nature of these inhibitions suggests some degree of overlap between amino acid entry through the three lysine sites. In the mammalian gut, by contrast, there are no such interactions between neutral and basic amino acids. Arginine transport by the culture form of *T. cruzi* is also atypical in its degree of specificity; neutral amino acids are competitive inhibitors of arginine transport, while the other basic amino acids are without effect. *Trypanosoma gambiense* transports a number of amino acids by saturable processes and there also appears to be a substantial diffusion component, predominant at higher amino acid concentrations.

Any study of solute transport in the Digenea is complicated by the presence of the gut. It is possible, however, to ligature the pharynx and thereby examine the uptake of amino acids and carbohydrates that occurs through the tegument, though in practice ligaturing has been found to be unnecessary since, in short-term studies, there is no difference in the rate of uptake with either the gut open or ligatured. *Schistosoma mansoni* uses a combination of mediated uptake and simple diffusion to absorb alanine, arginine, glutamate, glycine, methionine and tryptophan; proline enters exclusively by a saturable mediated system. There are indications that amino acid transport in schistosomes is unusual in that neutral amino acid transport is inhibited by basic amino acids. Additionally, there are quantitative changes in permeability to amino acids in *S. mansoni* during development: the free-swimming cercariae are virtually impermeable to methionine, recently penetrated schistosomula are somewhat more permeable, but by the time the worm has been domiciled in its host for three weeks the adult transport system has become developed.

By contrast with S. *mansoni*, amino acid transport in *Fasciola hepatica* takes place by simple diffusion only, an unusual situation deserving of further attention.

In *Hymenolepis diminuta* there are at least six kinetically distinguishable amino acid transport systems, the specificities of which vary widely. The four neutral amino acid systems show relatively little overlap with basic or acidic amino acids. On the other hand, some neutral amino acids interact with the acidic amino acid site, while the basic amino acid system has a very narrow specificity. All amino acid entry into *H. diminuta* occurs via mediated transport, although simple diffusion may also feature in the uptake of some, such as histidine and proline. These, along with α-aminoisobutyric acid and cycloleucine, are absorbed by active transport and are therefore accumulated against a concentration gradient. No such information is available for the other amino acids.

The complexity of amino acid transport of *H. diminuta in vivo* has been described by Read, who derived a formulation, based upon the Michaelis-Menten equation, to account for the transport of a given amino acid in the presence of a realistically large number of other amino acids (analogous to the situation pertaining in the mammalian gut):

$$V_1 = \cfrac{V_{max}}{\cfrac{K_t}{[S]} + 1 + \cfrac{K_t[S_1]}{K_{t_1}[S]} + \cfrac{K_t[S_2]}{K_{t_2}[S]} + \cdots + \cfrac{K_t[S_n]}{K_{t_n}[S]}}$$

where V_1 is the velocity of the uninhibited uptake of amino acid S, K_t is the Michaelis or transport constant, V_{max} is the maximum velocity of the uninhibited uptake and $[S_1]$, $[S_2]$ . . . $[S_n]$ are the concentrations of different inhibitory amino acids. This formulation has been tested by examining alteration in the uptake of methionine (V_1) by *H. diminuta* incubated in various mixtures of amino acids. The predicted value for V_1 is met by the observed value only when the K_t value is replaced with the K_i (inhibitor constant with respect to methionine) for each amino acid in the mixture. Good correlation is then obtained. The inhibitor constants (K_i) are numerical indications of the affinity of the inhibitory molecule for the transport site; they are determined by a variety of experimental procedures (see figure 2.7).

There are a number of important differences between amino acid transport systems of cestodes and those of mammals. For instance, mammalian systems maintain an almost exclusive stereospecific preference for the L-enantiomorph, whereas some tapeworms show an equal affinity for both D and L amino acids (*H. diminuta, H. citelli* and larval *Taenia crassiceps*). Another divergence is the apparent lack, in cestodes, of an ion (Na^+/Cl^-) requirement coupled to amino acid transport—most mammalian systems require ion coupling for efficient amino acid transport.

There is a paucity of data on amino acid uptake in the Acanthocephala. *Moniliformis dubius* and *Macracanthorhynchus hirudinaceus* both absorb amino acids by mediated systems and by simple diffusion. Leucine and methionine are actively transported and the site specificity of the methionine system (as judged by inhibitor studies) is narrow, being limited to neutral amino acids alone. As with the cestodes, there appears to be no sodium ion requirement for amino acid transport.

The majority of nematodes, as we have already noted, do not absorb nutrients across their external surfaces. There have been but a few studies on uptake of amino acids by the nematode intestine.

Lipid transport   This is a much neglected area of research in parasite nutrition and biochemistry. In general, parasites do not synthesize their long chain fatty acids *de novo* and therefore rely on the host to supply them with their basic needs. *H. diminuta* cannot synthesize long chain fatty acids but can lengthen the chain of absorbed fatty acids by one or two methylene units, by adding acetate units; hence fatty acid absorption becomes all-important. There are two distinct systems for fatty acid transport in *Hymenolepis diminuta*—one specific for short chain fatty acids ($C_2 - C_8$) and the other specific for longer chain acids ($C_{14} - C_{24}$). These two systems have been designated, somewhat arbitrarily, the "acetate site" and the "palmitate site" (table 2.2). Acetate is absorbed in *H. diminuta* at a rate four times greater than palmitate. Bile salts are required for long chain fatty acid transport but not for short chain transport. The role of bile acids may be indirect, since they are thought not to be absorbed by *H. diminuta* but instead become adsorbed to the tegument. Both of the fatty acid transport systems are specific to the fatty acids of appropriate chain length, and uptake is unaffected by any other class of compound. Cholesterol is absorbed by *H. diminuta* from solutions of mixed micelles with mono-olein and sodium taurocholate. The worm cannot synthesize cholesterol, a block residing at the step of

Table 2.2 Inhibitors and inhibitor constants (K_i) of fatty acid and nucleic acid precursor transport in the tapeworm *Hymenolepis diminuta*.

Substrate	Inhibitor	K_i (mM)
Short chain fatty acids		
acetate	formate	6.9
	propionate	1.3 (adult); 5.3 (cysticercoid)
	butyrate	3.6 („); 16 („)
	valerate	5.0
	octanoate	5.1
butyrate	acetate	5.4
Long chain fatty acids		
palmitate	pentadecanoate	4.9
	heptadecanoate	2.3
	stearate	2.3
	arachidate	3.5
	oleate	4.5
	linoleate	3.8
	linolenate	5.6
Nucleic acid precursors		
uracil	hypoxanthine	1.2
	thymine	-4.2*
hypoxanthine	adenine	6.0
adenine	hypoxanthine	1.0
uridine	adenosine	0.1
	AMP	0.1
	ATP	0.3

*Stimulation of uptake rather than inhibition
Data from Pappas and Read, (1975)

conversion of farnesol (farnesyl pyrophosphate) to squalene. The formation of micellar solutions by bile acids is clearly of importance to the efficient absorption of a number of compounds by *H. diminuta*, including free fatty acids, monoglycerides and cholesterol. On the other hand, glucose transport can be inhibited by sodium taurocholate. This is an area of some interest since it may in part explain the postulated role of bile salts in determining host specificity of certain parasites, though a great deal more data need to be collected before any assertions can be made.

Fatty acid transport in other groups of parasites has been virtually overlooked. *Fasciola hepatica* obtains short chain fatty acids (acetate and butyrate) by simple diffusion, but there is also an indication that acetate may enter by a mediated transport system shared by propionate, butyrate and valerate.

Purine and pyrimidine transport   Purine and pyrimidine uptake has been examined only in *H. diminuta*. The transport of these molecules is highly complex, compared with the other systems that we have discussed, in that there are at least three separate systems operating and two of these involve multiple binding sites. Additionally, inhibitory or stimulatory compounds can bind non-productively to a binding site, but are not themselves transported. Pappas, Uglem and Read proposed a model to account for the data obtained (figure 2.8). The thymine-uracil system transports thymine and uracil via an allosteric interaction and is thought to comprise a minimum of two binding sites. Non-productive binding of

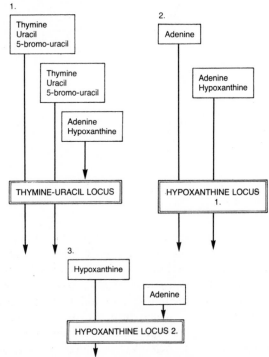

Figure 2.8 Proposed model for the transport of purines and pyrimidines in the cestode *Hymenolepis diminuta*. Where the arrows pass through the transport locus actual entry of the substrate occurs; arrows that do not pass through the locus indicate the occurrence of non-productive binding with no transport. Some solutes appear to bind to a single locus at two different sites. Based on Pappas, P. W. and Read, C. P. (1975).

adenine and hypoxanthine occurs at this site. There are two additional sites concerned with the transport of hypoxanthine, designated locus 1 and locus 2. Hypoxanthine, guanine and adenine are transported through locus 1 and adenine binds at two separate sites on the locus. Hypoxanthine is also transported via locus 2, which binds adenine non-productively. It is not clear why such a complex transporting system is found in *H. diminuta* and studies on purine and pyrimidine transport in other parasites would be of value in assessing general patterns.

Vitamin transport Very little experimental data on the vitamin requirements of parasites have been collected, let alone data on the modes of vitamin absorption. The B_{12} requirement of the broad tapeworm of man (*Diphyllobothrium latum*) is well known, but the mechanism of entry has not been examined. Thiamine and riboflavin enter *H. diminuta* by mediated transported processes with narrow specificity, while pyridoxine and nicotinamide enter by simple diffusion.

Macromolecular transport (*pinocytosis*) The uptake of macromolecules, such as proteins, across the surfaces of parasites has been the subject of only limited investigation. We can generalize by recording that many of the parasitic Protozoa obtain macromolecular nutrients by endocytosis, whereas helminth parasites do not seem to favour this process. There is, for example, no evidence for the endocytotic acquisition of the macromolecules ferritin and horseradish peroxidase by *Hymenolepis diminuta*, though both of these molecules are adsorbed onto the surface of the worm. Neither of the digeneans *Fasciola hepatica* and *Haematoloechus medioplexus* absorb colloidal material across the tegument, but the latter species acquires ferritin by endocytosis across the caecal gastrodermis. *Schistosoma mansoni*, on the other hand, absorbs horseradish peroxidase by endocytosis across the tegument.

Endocytosis has been examined in a number of species of parasitic Protozoa. The malaria parasite (*Plasmodium* sp.) feeds on host haemoglobin during the intraerythrocytic phase of the life cycle. The haemoglobin is absorbed intact through the cytostome by a process often referred to as *phagotrophy* (as opposed to osmotrophy, which is the absorption of low molecular weight molecules by simple diffusion or mediated transport). Phagotrophy, or endocytosis, in *Plasmodium* involves the formation of vesicles in which the haemoglobin is both absorbed and digested; these vesicles (or vacuoles) contain digestive enzymes and, ultimately, become filled with haem, the end-product of haemoglobin digestion. The formation of endocytotic vesicles takes place

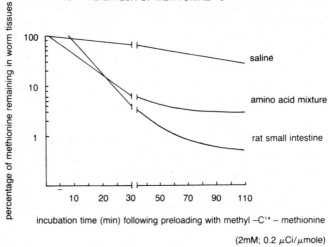

A. THE EFFLUX OF METHIONINE –C¹⁴

incubation time (min) following preloading with methyl –C¹⁴ – methionine

(2mM; 0.2 μCi/μmole)

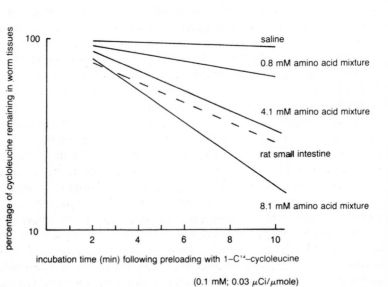

B. THE EFFLUX OF CYCLOLEUCINE –C¹⁴

incubation time (min) following preloading with 1–C¹⁴–cycloleucine

(0.1 mM; 0.03 μCi/μmole)

exclusively at the cytostome in *Plasmodium*. Other sporozoan parasites (gregarines and *Babesia rhodaini*) also have a cytostome. Among the parasitic amoebae, *Entamoeba histolytica* employs a surface lysosome to obtain the macromolecular contents and, presumably, dissolved nutrients, of the host cell. A cytostome is present in one group of haemoflagellates, the Kinetoplastida, but not in the Salivaria. Intact protein molecules are ingested in vesicles formed at the base of the flagellar invagination in the kinetoplastid trypanosomes and the cytostome itself may be involved. In *Crithidia fasciculata*, ferritin is absorbed in cytostomal vesicles. In the salivarian trypanosomes, on the other hand, macromolecules are absorbed only within vesicles formed in the flagellar invagination. Endocytosis may occur at localized regions of the body, such as the cytostome, when present, or over the entire surface of the parasite.

Bidirectional fluxes of nutrients

Movement of many nutrients, e.g. amino acids, can take place both into and out of parasites. The efflux of labelled methionine from *H. diminuta* is more rapid when worms are placed in a solution containing a mixture of amino acids resembling the proportions of amino acids found in the rat gut, than when incubated in a simple saline solution; the rate of efflux into the mixture is comparable to the rate of loss of labelled methionine in worms that are surgically implanted into rats (figure 2.9). This suggests the occurrence of counterflow mechanisms and demonstrates the mediated nature of the efflux process. Additional evidence for the dynamic balance that exists between the tapeworm and its host comes from the distribution of the radioactively labelled non-metabolizable amino acids, cycloleucine and α-aminoisobutyric acid, between rat and parasite following their injection. Both these amino acids become distributed in a steady state between host and parasite, whereas amino acid imbalance *in vivo*, induced by feeding a single amino acid, is rapidly

---

Figure 2.9 The kinetics of the *in vitro* and *in vivo* efflux of amino acids from the cestode *Hymenolepis diminuta*. In both of these experiments, tapeworms were incubated in radiolabelled amino acid and the subsequent efflux of the preloaded acid was determined either *in vitro* (in saline or mixtures of amino acids in the ratios found in the small intestine of the rat) or *in vivo* following surgical transplantation of the worms into the rat gut.
A. redrawn from Hopkins, C. A. and Callow, L. L. (1965) *Parasitology*, **55**, 653–666.
B. redrawn from Arme, C. and Read, C. P. (1969) *Comparative Biochemistry and Physiology*, **29**, 1135–1147.

RAT SMALL INTESTINE (LUMEN CONTENTS)

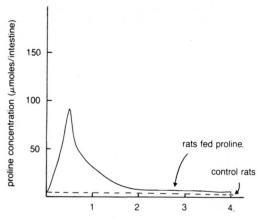

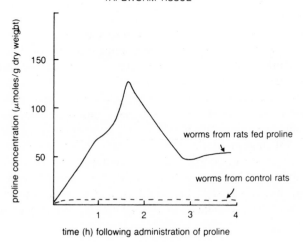

time (h) following administration of proline

Figure 2.10 The differential handling of an experimentally-induced imbalance of a single imino acid (proline) by the rat and its tapeworms (*Hymenolepis diminuta*). Rats, infected with 30 worms each, were fed a single dose of 250 mg proline; the levels of proline in the rat gut and in the tapeworm tissue were determined by amino acid analysis at intervals up to 4h post-feeding.

Redrawn from Chappell, L. H. and Read, C. P. (1973) *Parasitology*, **67**, 289–305.

Table 2.3 Kinetics of membrane transport in the digenean, *Schistosoma mansoni.*

Substrate	K_m (mM)	V_{max}(μmol g^{-1} h^{-1})
Carbohydrates		
glucose	0.8	147
galactose	0.2	10
2-deoxyglucose	1.7	270
Amino acids		
methionine	0.3–1.9	120–192
glycine	0.6–1.1	165–225
proline	1.7–2.0	750
arginine	0.1–0.4	45–60
glutamic acid	0.3–0.5	75
tryptophan	0.5–1.0	90–180
Nucleic acid precursors		
hypoxanthine	0.9	45
adenosine	0.5	52
uridine	0.4	22
adenine	0.05	10

Data from Pappas and Read, (1975)

restored in the host but not in the worm (figure 2.10). This may indicate that tapeworms are deficient in regulatory mechanisms and therefore, to some extent, parasitize the homoiostatic mechanisms of their hosts. Despite this notion, there is evidence that amino acid imbalances may have little or no effect upon worm protein synthesis and overall growth. The notion of parasites existing in dynamic equilibrium with their hosts is an exciting one that has been sadly ignored.

Surface enzymes in parasites

Enzymes bound to the surfaces of parasites may be of parasite origin (intrinsic) or derived from their hosts (extrinsic).

Intrinsic surface enzymes

Enzymes of parasite origin may play a role in digestion, as well as in penetration and migration through host tissues. We shall discuss the latter enzymes in a later chapter. Cestodes were long thought to rely exclusively upon the digestive enzymes of their hosts to provide them with nutrient molecules in a state available for absorption, but recent work has shown that tapeworms may possess their own digestive enzymes bound to their surface. Phosphohydrolase activity is present in the tegument of *H. diminuta*, and its action permits the absorption of otherwise impermeable phosphate esters following hydrolysis, e.g. fructose 1, 6-diphosphate is hydrolysed at the surface to yield inorganic phosphate that can then be absorbed; fructose itself is unable to enter *H. diminuta*. The spatial proximity of the glucose transport system in *H. diminuta* to membrane-bound phosphohydrolases could confer a kinetic advantage to the absorption of carbohydrates. The same may be said of the ribonuclease activity associated with the tegument of *H. diminuta* and the hydrolysis of RNA to yield nucleosides that can then be acquired by the parasite.

Extrinsic surface enzymes

Ugolev coined the terms *membrane* or *contact digestion* to account for the enhancement of starch digestion by pancreatic α-amylase in the presence of isolated segments of vertebrate intestine. It was proposed that the enzyme became adsorbed on to the glycocalyx of the mucosal cell surface, in which condition amylolytic activity was increased. Contact digestion also occurs with tapeworms, since the activity of pancreatic α-amylase is also enhanced *in vitro* by the presence of living tapeworm tissue (*H. diminuta*, *H. microstoma* and *Moniezia expansa*). Enhancement does not occur if the tapeworms are killed with ethanol. None of these worms have intrinsic amylolytic activity and it is therefore assumed that, *in vivo*, enzymes of host origin may become bound to the glycocalyx of the worms, providing the spatial advantage of proximity of the products of enzyme activity to the worm surface. The acantho-cephalan *Moniliformis dubius* can bind host amylase, but not trypsin, chymotrypsin or lipase.

Inhibition of host enzymes by parasites

Antienzymes, with an inhibitory effect upon their host's enzymes, are reported to be common amongst intestinal parasites. Antienzymes from ascarids are inhibitors of trypsin or pepsin. *Ascaris suum* is thought to possess an antitrypsin and an antichymotrypsin located in its cuticle, intestine and other tissues. *Hymenolepis diminuta* possesses a number of antienzymic properties, e.g. trypsin, α-chymotrypsin and β-chymotrypsin are irreversibly inactivated *in vitro* and pancreatic lipase undergoes a reversible inhibition. No antienzyme as such has been isolated from *H. diminuta* and it has been suggested that these enzyme inactivations take place within the glycocalyx of the worm. Their precise mechanism, however, is unknown. While there are obvious advantages to a parasite that inhabits the upper intestine of vertebrates in having the ability to inhibit host digestive enzymes, the high rate of turnover of the parasite surface may render antienzymic properties redundant.

SUMMARY

1. *In vitro* culture of parasites is a necessary tool to aid the understanding of the nutritional requirements of parasites, but successful growth in chemically defined media has been achieved with only a handful of parasites.

2. An alimentary tract is present in some parasites (Monogenea, Digenea and Nematoda) but is absent from others (Protozoa, Cestoda and Acanthocephala). The morphology of the parasite gut is consistent with a digestive/absorptive function, but there is inadequate information to establish the precise role of this organ.

3. There is considerable morphological and biochemical evidence that the external surfaces of most parasites play a fundamental role in the acquisition of nutrients. The complexity of the mechanisms involved is gradually being revealed, but only a few species have been studied in detail.

4. The absorptive mechanisms employed by parasites to obtain nutrients across the surface membranes include simple diffusion, facilitated diffusion and active transport. The specific properties of these transport systems may differ considerably from mammalian systems.

5. Endocytosis is a common route of entry for macromolecules into protozoan parasites. Helminths only rarely use this process.

6. Some gut parasites can bind host enzymes to their surface membranes and cause either enhancement or inhibition of enzyme activity. Antienzymes have been isolated from a few gut parasites. Intrinsic digestive enzymes may be released at the surfaces of parasites.

CHAPTER THREE

CARBOHYDRATE METABOLISM
AND ENERGY PRODUCTION

Stored carbohydrates

Polysaccharide An almost universal feature of endoparasitic organisms is their dependence upon anaerobic carbohydrate metabolism to obtain energy, regardless of the amount of environmental oxygen available. Accordingly, a great many parasites store polysaccharides which can be oxidised (though rarely to completion) to yield ATP. The most common polysaccharide reserve is glycogen, occurring in protozoans and helminths alike. Glycogen is composed exclusively of glucose molecules joined by α-1, 4 and α-1, 6 linkages, the ratio of 1, 4:1, 6 bonds ranging from 12:1 to 15:1.

The glycogen content of parasites varies quantitatively according to species. Only 1% of the dry body weight of fish gill monogeneans is glycogen, whereas higher amounts are present in the tissues of internal parasites. Digeneans, for example, may store between 2% and 30% of their dry weight as glycogen, and cestodes tend to contain even greater quantities (20–60% of dry weight). Nematodes and acanthocephalans also contain relatively large amounts of stored glycogen, varying between 10 and 60% of their dry weight. Even within a single species, sexual dimorphism of stored glycogen content may occur, e.g. male schistosomes store much more glycogen (14–30% of dry weight) than do female worms (3–5% of dry weight). The metabolic significance of this difference is unclear. Sexual difference in the quantities of stored glycogen is also a feature of several other parasites: *Macracanthorhynchus hirudinaceus, Moniliformis dubius* (Acanthocephala); *Ascaris lumbricoides, Ascaridia galli, Parascaris equorum* and *Angiostrongylus cantonensis* (Nematoda).

Utilization of glycogen as an energy store can be demonstrated readily by experimental deprivation of either carbohydrate or total food of the infected host, whereupon rapid depletion of stored glycogen within the parasite occurs. The chicken tapeworm, *Raillietina cesticillus* reduces its glycogen store from approximately 25% of its dry weight to 1.5% in just 24 hours if the host chicken is denied access to food. *Hymenolepis diminuta* shows a similarly dramatic glycogen loss if the host rat is starved of carbohydrate. Glycogenesis, the reverse process, takes place rapidly if experimentally starved worms are then given access, either *in vitro* or *in vivo*, to a utilizable glycogenic substrate (monosaccharide, pyruvate, citrate, succinate, glutamate, isoleucine, valine or glyoxylate, depending upon the parasite species).

Helminth parasites store glycogen in their parenchymal cells, but significant quantities also occur in the musculature. Generally speaking, however, determinations of the glycogen content of parasites are made on extracts of the whole body and no attempt is made to distinguish between the various tissues. A feature of glycogen determination, presumably pertaining to all animal tissues, is that the molecular weights of extracted glycogen vary quite widely according to the extraction procedures used, e.g. aqueous extraction of glycogen from *Fasciola hepatica* yields a broad spectrum of different molecular weight moieties, whereas most of the higher molecular weight glycogen is lost using potassium hydroxide as an extractant. The terms lyo- and desmo-glycogen, referred to in some studies, are now considered erroneous, the different forms representing artifacts of the extraction procedures.

Some of the parasitic Protozoa store glycogen, but an alternative polysaccharide reserve, amylopectin—a highly branched polymer—occurs in many species.

Simple sugars   The presence of simple sugars has been detected in many parasites, but these tend to occur at rather low concentrations. Glucose, fructose, galactose, mannose, xylose and ribose are monosaccharides that occur most frequently. The disaccharide trehalose, most abundant in the eggs of *Ascaris lumbricoides*, is a common constituent of protozoan and helminth tissues. Nematodes may also contain glycosides, whose sugar component is a unique hexose, ascarylose (3, 6-dideoxy-L-arabino hexose). These glycosides, termed *ascarosides*, occur in the egg shells of a number of nematodes, including *Ascaris lumbricoides*, *Parascaris equorum*, *Ascaridia galli* and *Toxocara cati*.

Glycolysis

Glycolysis is a fundamental pathway of carbohydrate catabolism common to almost all animal tissues. It is sometimes referred to as the

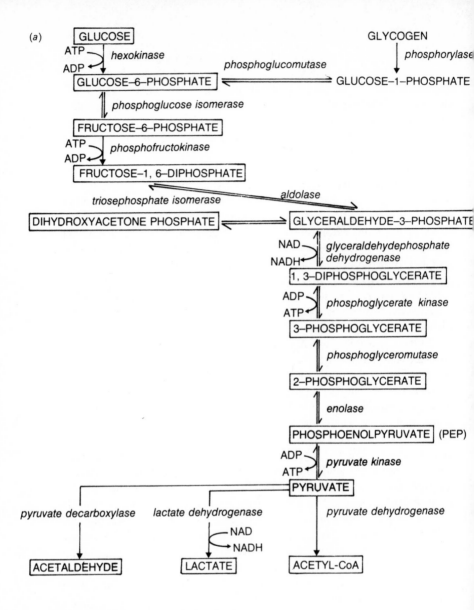

Figure 3.1 (*a*) The glycolytic pathway of carbohydrate catabolism. (*b*) Net reaction of classical glycolysis.

Embden-Meyerhof-Parnas pathway after those workers who, in the early decades of this century, described the breakdown of glucose to pyruvate under anaerobic conditions. Glycolysis is used by free-living animals as a relatively rapid means of obtaining energy, usually for muscular contraction, from stored glycogen. During this process an "oxygen debt" is accumulated which must be "repaid" under aerobic conditions, in order to oxidise the reduced products of glycolysis, such as lactate, whose accumulation becomes toxic. During repayment of the oxygen debt, part of the lactate is reoxidised to provide energy for glycogen synthesis from the remainder. Glycolysis, therefore, is a short-term pathway in free-living animals, e.g. in teleost fish white muscle, which is anaerobic and can only be used for short bursts of rapid swimming. Parasites, by contrast, may rely on anaerobic metabolism to a much greater extent than most free-living animals. Many internal parasites, particularly those inhabiting various regions of the vertebrate and invertebrate alimentary canal, live under conditions of greatly reduced oxygen tension. As a consequence we find that these parasites often rely on glycolysis as a major route of carbohydrate catabolism, rather than as an emergency source of energy. The net result of this is that these parasites only partially oxidise substrates such as hexose sugars, and excrete reduced energy-rich products, which, perhaps, may then be used by their hosts.

The glycolytic pathway has been examined in a wide variety of parasites, and, in most cases, a typical array of enzymes and intermediate metabolites is found (figure 3.1). i.e. parasite glycolysis appears to function as it does in other, non-parasitic, animals. However, in some parasites various enzymes of the glycolytic pathway have not been detected, but whether this alone is sufficient evidence to indicate that glycolysis is an unimportant source of energy in these parasites is a matter for conjecture.

End-products of carbohydrate catabolism

The typical end-product of glycolysis is lactic acid. The reduction of pyruvate to lactate serves to reoxidise the reduced cofactor NADH produced by glyceraldehyde-3-phosphate dehydrogenase (figure 3.1). Despite this general statement, parasite catabolism is typified by the production of a wide variety of end-products—many more than are produced by vertebrates or micro-organisms.

Parasites that are homolactic fermenters, such as *Schistosoma mansoni*, *Brugia pahangi* and *Dipetalonema viteae*, excrete lactic acid almost exclusively, under both anaerobic and aerobic conditions. Many other parasites excrete volatile fatty acids—trypanosomes excrete acetate, and some ciliates excrete acetate, propionate and butyrate aerobically. Among the helminths examined, *Fasciola hepatica*, *Moniliformis dubius*,

Table 3.1 The end-products of carbohydrate catabolism in selected parasites.

Species	Conditions	End-products
Protozoa		
Crithidia fasciculata	aerobic	pyruvate, succinate, ethanol, CO_2.
	anaerobic	lactate, succinate, ethanol, CO_2.
Trypanosoma brucei	aerobic	pyruvate, acetate, succinate, glycerol, CO_2.
Trypanosoma lewisi	aerobic	lactate, pyruvate, acetate, succinate, glycerol, ethanol.
	anaerobic	lactate, pyruvate, acetate, succinate, ethanol.
Trypanosoma gambiense	aerobic	acetate, pyruvate, CO_2.
	anaerobic	lactate, pyruvate, acetate, succinate, CO_2.
Plasmodium berghei	aerobic	lactate, pyruvate, acetate, succinate, CO_2.
Digenea		
Fasciola hepatica	aerobic	lactate, acetate, propionate, α-methylbutyrate, α-methylvalerate.
	anaerobic	lactate, acetate, propionate, α-methylbutyrate, α-methylvalerate, succinate.
Cestoda		
Hymenolepis diminuta	aerobic	lactate, acetate, succinate.
	anaerobic	lactate.
Acanthocephala		
Moniliformis dubius	aerobic	lactate, formate, acetate, propionate, α-methylbutyrate, succinate, ethanol.
Nematoda		
Ascaris lumbricoides	aerobic	lactate, formate, α-methylbutyrate, α-methylvalerate, C6 acids, succinate.
	anaerobic	lactate, formate, acetate, propionate, α-methylbutyrate, succinate, acetone.

Data from Von Brand, (1973)

Ascaris lumbricoides and *Trichinella spiralis* excrete a variety of volatile fatty acids, including acetate, propionate, butyrate and, to a lesser extent, valerate, isovalerate, isobutyrate and caproate. The proportions of fatty acids produced varies with the prevailing conditions, e.g. *Fasciola* excretes less propionate and proportionally more acetate with increasing oxygen tension. Ethanol is an end-product of carbohydrate catabolism in some parasites (larval scolices of *Echinococcus granulosus*, larval and adult *Taenia taeniaeformis* and *Moniliformis dubius*.) This is reminiscent of alcoholic fermentation found in the yeasts. Details of the end-products of carbohydrate catabolism are given in table 3.1.

Glycolytic enzymes of parasites

There is insufficient space to consider in detail the characteristics of parasite glycolytic enzymes and readers are directed to the comprehensive account by Von Brand (1973) for further reading. Suffice it here to indicate areas of particular interest concerning parasite glycolysis.

Hexokinases are a group of enzymes concerned with phosphorylating hexoses and thereby allowing their entry to the glycolytic pathway. *Schistosoma mansoni* and *Echinococcus granulosus* contain a number of different specific hexokinases, while trypanosomes possess a single hexokinase with a broad spectrum of phosphorylating activity. In general, cestodes metabolize a limited number of sugars, usually only glucose and galactose. Interestingly, few parasites metabolize sugars other than monosaccharides, implying the absence of disaccharidases from parasite tissues. Noteable exceptions to this generalization are *Moniliformis dubius* and *Ascaris lumbricoides*. It should be recognized that the inability of a particular parasite to utilize a given sugar may not necessarily reflect the absence of the appropriate hexokinase, e.g. while many parasitic flagellates and some helminths metabolize fructose, the cestode *Hymenolepis diminuta* cannot, due to its impermeability to that hexose rather than any enzyme deficiency.

Partial purification of parasite glycolytic enzymes reveals the occurrence of several forms (isoenzymes) of a single enzyme. Lactate dehydrogenase (converting pyruvate to lactate) may occur in from one to five electrophoretically different forms depending upon parasite species. Although it is often difficult to ascribe a function for several physico-chemically different enzymes that catalyze the same chemical conversion, it may be that these isoenzymes variously function at different stages of the parasite life-cycle, or only in certain tissues.

Carbon dioxide fixation

Many parasites, particularly gut dwellers, are exposed to environmental conditions with a characteristically high CO_2 tension. It is no surprise, therefore, to find that CO_2 can be fixed via certain well described metabolic routes, serving both a biosynthetic and energetic function. The presence of radiocarbon derived from $C^{14}O_2$, in both the end-products of carbohydrate catabolism and in stored polysaccharide, indicates the overall direction of CO_2 fixation. Current research has brought to light some of the specific pathways used for CO_2 fixation by parasites. Of the known enzyme systems capable of CO_2 fixation, two occur in parasites. These involve either "malic enzyme" (ME) or phosphoenolpyruvate carboxykinase (PEPCK).

"Malic enzyme" occurs in a variety of forms, one of which (L-malate: NAD oxidoreductase (decarboxylating) E.C. 1.1.1.39) is found exclusively in *Ascaris lumbricoides* and catalyses the carboxylation of pyruvate to form malate. This enzyme is NAD-dependent and the reaction it catalyzes is reversible. Another form of the enzyme (L-malate: NADP oxidoreductase (decarboxylating) E.C. 1.1.1.40), which is found in many animal tissues, is dependent upon the cofactor NADP. Phosphoenolpyruvate carboxykinase, now usually referred to as PEP carboxylase, carboxylates pyruvate to form oxaloacetate and is dependent upon either guanosine triphosphate (GTP) or inosine triphosphate (ITP). The distribution of ME and PEPCK in parasites is shown in table 3.2, which effectively illustrates the paucity of data currently available. In fact, the details of the mechanisms of CO_2 fixation are derived from studies on a very small number of parasites indeed.

Radiocarbon from $C^{14}O_2$ can be detected in both glycogen and end-products of catabolism, such as pyruvate and succinate, in isolated *Ascaris* muscle. The proposed pathway for CO_2 fixation in this system is shown in figure 3.2. The reaction scheme places "malic enzyme" within the mitochondria, and PEP carboxykinase in the cytosol of the cell. Malate is formed from oxaloacetate (catalyzed by NADH-dependent malate dehydrogenase) and diffuses across the mitochondrial membrane, where a dismutation reaction is thought to occur, cleaving malate to pyruvate and fumarate, by the respective actions of "malic enzyme" and fumarate hydratase. The NADH generated by "malic enzyme" is reoxidized during the reduction of fumarate to succinate, at which step one mole of ATP is produced per mole of fumarate reduced. Typically, parasites capable of CO_2 fixation excrete succinate and pyruvate, but each of these may be

Table 3.2 Occurrence of "malic enzyme" (ME) and PEP carboxykinase (PEPCK) in some helminth parasites.

Parasite species	ME	PEPCK
Digenea		
Dicrocoelium dendriticum	?	+
Fasciola hepatica	+	+
Schistosoma mansoni	?	+
Cestoda		
Echinococcus granulosus	+	+
Hymenolepis diminuta	+	+
Moniezia expansa	+	+
Acanthocephala		
Moniliformis dubius	+ (?)	+
Nematoda		
Ascaris lumbricoides	+	+
Dictyocaulus viviparus	−	+
Haemonchus contortus	+	+
Nippostrongylus brasiliensis	−	+
Syphacea muris	−	+
Trichinella spiralis	+	+

+ = present
− = not detected
(?) = not determined or uncertain presence
Data from Bryant (1975)

further metabolized, succinate to volatile fatty acids and pyruvate to lactate, alanine or volatile fatty acids.

The cestode *Hymenolepis diminuta* excretes succinate and there is good evidence that, like *Ascaris*, it also fixes CO_2 via "malic enzyme" and PEP carboxykinase, to yield succinate and synthesize limited amounts of glycogen. The reactions of CO_2 fixation in *Hymenolepis* are essentially identical to those of *Ascaris*, with the exception of the occurrence of a transhydrogenase, responsible for reducing NAD in the presence of NADPH in the former species. This is essential for redox balance, since the "malic enzyme" of *Hymenolepis* is NADP-dependent while that of *Ascaris* is NAD-dependent.

The advantage of these mixed fermentations that occur in many parasites may be considered as threefold. First, the decarboxylations of succinate to propionate, and pyruvate to acetate will yield potential metabolic energy, since both propionyl-CoA and acetyl-CoA are high

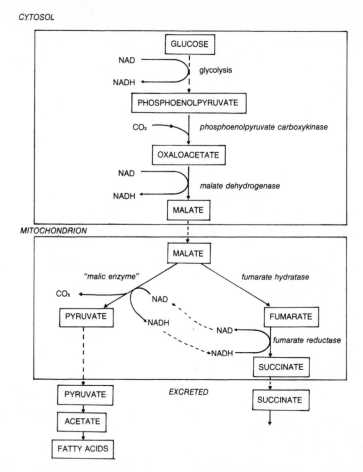

Figure 3.2   The roles of "malic enzyme" (ME) and phosphoenolpyruvate carboxykinase (PEPCK) in CO_2 fixation by the nematode *Ascaris lumbricoides* (from Bryant, 1975).

energy compounds. Secondly, energy budgets are such that homolactic fermentations, such as occur in *Schistosoma*, yield only 2 moles of ATP per mole of glucose oxidised; this yield increases by one mole of ATP if CO_2 is fixed, as in *Ascaris* and *Hymenolepis*. Also 2 moles of ATP will be obtained if pyruvate and succinate are decarboxylated to the appropriate end-product. Mixed fermentations generate and utilize reducing equivalents (NAD↔NADH, NADP↔NADPH) allowing for some degree of flexibility in balancing reactions coupled to redox potentials. The third

Table 3.3 Ratios of the activities of pyruvate kinase (PK) to phosphoenolpyruvate carboxykinase (PEPCK) in selected helminths.

Parasite	PK : PEPCK	Major end-product
Digenea		
Dicrocoelium dendriticum	0.05	lactate, propionate, succinate, acetate
Schistosoma mansoni (adult) male	5.0	lactate
female	9.7	lactate
Cestoda		
Hymenolepis diminuta (adult)	0.18	succinate, lactate
Nematoda		
Ascaris lumbricoides (adult)	0.04	succinate, volatile fatty acids
Nippostrongylus brasiliensis (adult)	1.86	succinate, lactate
Dirofilaria immitis (adult)	10.6	lactate (?)
Acanthocephala		
Moniliformis dubius (adult)	0.34	lactate, succinate
(larvae)	—	lactate, succinate

Data from Bryant, (1975)

possible advantage of mixed fermentations is related to volatile fatty acid excretion; these end-products are likely to be less damaging to the host tissues due to their low pKa values.

Regulation of carbohydrate catabolism

The ratio of the activities of two key enzymes in carbohydrate catabolism, pyruvate kinase (PK) and PEP carboxykinase, is responsible, in part, for the nature of the metabolic end-products excreted. High PK activity will lead to the formation of lactate from PEP and high PEPCK activity will lead to the production of succinate or its metabolites. Here we can see that PEP forms a major branch point in carbohydrate metabolism. As expected, homolactic fermenters, like schistosomes, have a high ratio of these enzymes, while succinate producers have a much lower value (table 3.3). One current hypothesis proposes that control of the direction of metabolism from PEP to either succinate or lactate, in gut parasites, is mediated through the partial pressure of CO_2 in the host's intestinal tissue.

The tricarboxylic acid cycle

Complete oxidation of carbohydrates, and other metabolites, such as amino acids and fatty acids, is carried out aerobically via the Tricarboxylic Acid (TCA) Cycle, also referred to as the Krebs or Citric Acid Cycle (figure 3.3). The cycle functions both catabolically to yield ATP, and biosynthetically as a fundamental source of carbohydrates, proteins and lipids.

The classical methodology for examination of the TCA cycle (isolation

(a)

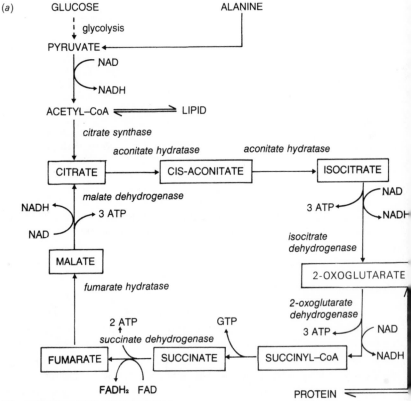

(b)   NET REACTION OF THE CYCLE:

Acetyl–CoA + 3NAD$^+$ + FAD + GDP + P + 2H$_2$O → 2CO$_2$ + CoASH + 3NADH + FADH$_2$ + GTP

11 moles of ATP are produced by the TCA cycle when it leads to electron transport and oxidative phosphorylation.

Figure 3.3 (a) The tricarboxylic acid cycle (citric acid or Krebs cycle).
(b) Net reaction of the cycle. 11 moles of ATP are produced by the tricarboxylic acid cycle when it leads to electron transport and oxidative phosphorylation.

and purification of the component enzymes, isolation and identification of intermediate products, radiocarbon distribution in intermediates of the cycle and in end-products, and the use of specific inhibitors and stimulatory molecules) has been applied to many parasite tissues. Nevertheless, the data available on the TCA cycle in parasites are incomplete and restricted to a small number of species. We can, however, tentatively make some general statements. Some parasites, living under conditions of relatively high oxygen tension, may possess a fully functional cycle, while others appear to possess only some of the cycle enzymes and intermediates. A small number of parasites, on the other hand, do not have a demonstrably functional TCA cycle at all. Because our knowledge of the TCA cycle in a sufficient number of different parasites is lacking, it is not possible to make meaningful generalizations on the presence or absence of the cycle relative to parasite habitat. Even the implication of a non-functional cycle may be due to the failure to demonstrate a particular enzyme rather than to its actual absence.

Organisms with a fully operational TCA cycle include avian malarial parasites, some species of trypanosomes (particularly forms grown in culture) and *Trichomonas gallinae*, among the Protozoa, and larval schistosomes, along with many nematodes, in the helminths. The absence of the cycle has been established unequivocally in *Hymenolepis diminuta*, *Moniezia expansa*, *Fasciola hepatica* and *Ascaris lumbricoides* (adults but not eggs). In these species there can be little doubt that the cycle makes no significant contribution to the energy budget of the parasite, though it has been suggested that a vestigial pathway could have a biosynthetic function. Even though the complete cycle is absent in these parasites, some species (*Hymenolepis* and *Ascaris*) utilize part of the cycle in the reverse direction, to reduce malate to fumarate and thence to succinate; this would appear to be a conservative adaptation to the special conditions of partial anaerobiosis experienced, particularly, by gut-dwelling parasites. It is worth noting that similar biochemical adaptations may occur in free-living animals whose environment is characterized by periods of oxygen depletion, e.g. intertidal molluscs.

The role of oxygen in parasite energy metabolism

Our discussion of energy metabolism, so far, leads us to the inevitable conclusion that many parasites obtain their energy from anaerobic pathways; such parasites tend to inhabit regions of low oxygen tension, such as the lumen of the vertebrate alimentary canal. Yet almost all parasite species do consume oxygen. Furthermore, metabolic end-products frequently vary according to whether aerobic or anaerobic conditions prevail (see table 3.1.). This, then, raises the important question of the role of oxygen in organisms whose metabolism is primarily anaerobic. Some oxygen will undoubtedly be used biosynthetically, e.g. in egg-shell synthesis, nevertheless most workers are agreed that oxygen also plays a significant role in the energy metabolism of many parasites.

The key questions would appear to be:

1. What are the mechanisms of oxygen uptake?
2. Does oxygen uptake lead to the production of ATP?
3. If this latter process (i.e. oxidative phosphorylation) does occur, what overall contribution does it make to the energy budget of the parasite?

Some of the better studied large helminths are unusual in that they are oxygen conformers, the rate of oxygen uptake depending upon its partial pressure. Hydrogen peroxide is often formed from oxidative metabolism, and cyanide, normally an inhibitor of oxidative respiration, is either without effect or it may even stimulate oxygen uptake. These observations point to the occurrence of flavoprotein oxidases rather than a conventional cytochrome system, but the latter, as we shall discuss shortly, does function in these worms.

We must not loose sight of the fact that much of the data obtained for parasite energy metabolism derives from experimental conditions containing high oxygen tensions, such as those of air (160 mm Hg)—the parasite in its natural habitat may be confronted with much lower oxygen tensions than this.

Considerable controversy surrounds the role of oxygen in parasite energy metabolism. One view is that many of the gut-dwelling nematodes, and possibly other helminths, are in fact aerobic in their natural habitat, despite the apparent importance of glycolysis discussed earlier. The argument continues by noting that the anaerobic metabolism determined *in vitro* may well be erroneous, due, in part, to the fact that worms removed from their hosts may rapidly become moribund, and data are therefore obtained from unhealthy rather than "normal" tissue. Unfortunately, satisfactory procedures for the examination of parasite metabolism *in vivo* have yet to be developed. Nevertheless, while Smith's (1969) cogent line of argument urges caution in the interpretation of metabolic data obtained from *in vitro* studies and should not be ignored, much of the current evidence strongly suggests that helminth parasites rely to varying extents on anaerobic metabolism, though many species

suitable environmental conditions. Very few parasites are actually intolerant of oxygen; some rumen ciliates, flagellates from the gut of termites, several trichomonad species and adult *Ascaris* appear to find high oxygen tensions inimical.
have the capability to switch to normal aerobic metabolism under

Pasteur and Crabtree effects

In most animal tissues that degrade glucose to pyruvate via glycolysis, the rate of fermentation is more rapid anaerobically than aerobically. This inhibition of glycolysis by oxygen is termed the *Pasteur Effect*. Its converse is the *Crabtree Effect*, whereby oxygen consumption is inhibited by glucose.

The fact that glycolysis in many helminth parasites persists in the presence of high oxygen tensions (i.e. limited or no Pasteur Effect) lends support to the view that anaerobiosis is indeed fundamental to their energy metabolism. A small number of species do exhibit a Pasteur Effect including larval *Schistosoma mansoni*, some larval nematodes such as *Ascaris lumbricoides*, *Eustrongyloides ignotus*, *Litomosoides carinii* and *Nippostrongylus brasiliensis*. The Crabtree Effect is regarded as a likely occurrence in those parasites with a demonstrably active glycolytic pathway.

Electron transport and terminal oxidations

Oxidative phosphorylation (aerobic ATP synthesis) takes place within mitochondria by means of a sequence of electron acceptors and donors (electron transport). The sequence commences at NAD, and ends with oxygen, which acts as the final electron acceptor. Between these two lie a chain of flavoproteins, including flavin mononucleotide (FMN) and flavin adenine-dinucleotide (FAD), and the cytochromes. The latter, in mammals, are designated b, c_1, c, a and a_3 according to their absorption characteristics rather than their sequence in the cytochrome chain. Interestingly, the pioneering studies of Keilin (1925) on animal respiratory systems and the discovery of the cytochromes included observations of the nematode *Ascaris*. Quinone coenzymes, coenzyme Q (CoQ) or ubiquinone (UQ), are also characteristic components of respiratory systems. A series of enzymes, flavoprotein and cytochrome oxidases, catalyze the transfer of electrons along the respiratory chain.

The ever-growing literature on electron transfer and terminal oxidation in parasites is complicated, and space available does not permit a detailed account here.

The cytochrome system can operate at oxygen tensions as low as 5 mm Hg and, it is, therefore, possible for an active system to function in many parasites, even vertebrate gut inhabitants. In the latter, as we have already discussed, CO_2 fixation and the reduction of fumarate to succinate by fumarate reductase, along with the excretion of a variety of reduced end-products, is an important feature of metabolism. Nevertheless, helminths such as *Fasciola hepatica*, *Hymenolepis diminuta*, *Moniliformis dubius* and *Ascaris lumbricoides* possess a functional

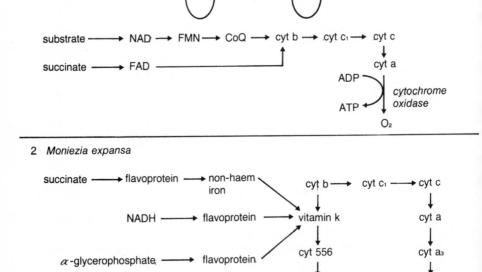

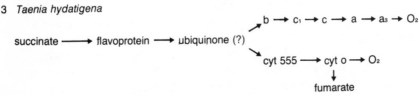

Figure 3.4   Electron transport chains.

cytochrome chain. It differs from the classical mammalian cytochrome system (figure 3.4) by being branched and possessing multiple terminal oxidases. One branch resembles the mammalian system by terminating in cytochrome a_3 and cytochrome oxidase, while other branches may terminate in cytochrome o (*Fasciola* and *Moniezia*) or, in addition, cytochrome a_1 (*Ascaris*)—the latter may be due to bacterial contamination, however. These parasite cytochrome systems are capable of oxidative phosphorylation apparently only through the mammalian-type branch; none occurs through the branch terminating in cytochrome o. The mechanisms responsible for controlling the direction of the flow through such branched cytochrome systems remain to be elucidated. Cytochrome o may be the source of hydrogen peroxide in some helminths and a number of peroxidases have been identified. Hydrogen peroxide may, additionally, be produced by other cytochrome oxidases or by flavoprotein oxidases.

The P : O ratio, a measure of oxidative phosphorylation efficiency, is three for the classical mammalian cytochrome chain, but it is reduced if one of the alternative parasite pathways is used. P : O ratios of one will be obtained if NADH is oxidized via cytochrome o. Branched cytochrome chains are not unique to helminth parasites and they are also found in many micro-organisms as well as parasitic Protozoa.

The succinoxidase system of *Ascaris*, involving succinic dehydrogenase and a terminal respiratory chain, has been examined in some detail. This nematode contains the classical mammalian-type cytochrome chain but produces hydrogen peroxide from the oxidation of succinate. Furthermore, cytochrome inhibitors such as azide, cyanide and Antimycin A, have little or no effect upon respiration. If the cytochrome o system of *Ascaris* plays a role in the reduction of fumarate to succinate, while at the same time the mammalian-type chain can operate normally under certain conditions, then respiration in this helminth is typical of that of a facultative, rather than obligate, anaerobe. This situation may well pertain to other helminths in which there are both normal cytochrome chains and alternative branched pathways.

Cytochromes and terminal oxidations have been examined in a limited number of protozoan parasites, with trypanosomes, trichomonads, and malarial parasites attracting the greatest attention. The cytochrome systems of some trypanosome species are sensitive to azide and cyanide (e.g. *Trypanosoma cruzi*), whereas partial or complete insensitivity to these inhibitors of the classical cytochrome system is demonstrable in other species (*T. vivax*, *T. congolense*, *T. rhodesiense*). One of the more

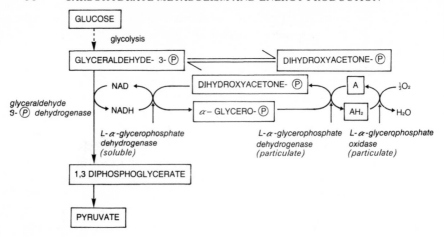

Figure 3.5 Proposed L-α-glycerophosphate oxidase pathway of trypanosomes (from Hill, 1976).
$\circledP$ = phosphate, A = unknown component

interesting features of trypanosome electron transport systems resides in the operation of different systems at various stages of the parasite life-cycle. This applies particularly to the African trypanosomes (*T. brucei* and *T. rhodesiense*) which have morphologically and biochemically distinct forms living in the vertebrate blood-stream and the insect midgut—the culture form of the parasites morphologically resembles the insect midgut form. The bloodstream trypanosomes possess an α-glycerophosphate oxidase system, oxidising L-α-glycerophosphate to dihydroxyacetone phosphate, using oxygen as the terminal acceptor (figure 3.5). The prominent feature of respiration in culture forms is the partial sensitivity to cyanide. The L-α-glycerophosphate oxidase system may account for as much as 95% of respiration in bloodstream trypanosomes, whereas additional oxidases are produced on transformation of bloodstream forms in culture. These culture-form oxidases have a higher affinity for oxygen than the α-glycerophosphate oxidase system, a feature possibly related to the lower oxygen tension of the insect midget compared with the mammalian bloodstream.

Respiratory pigments in parasites

Haemoglobin occurs in a small number of parasites, notably digeneans and nematodes. Interesting recent studies on the haemoglobins in

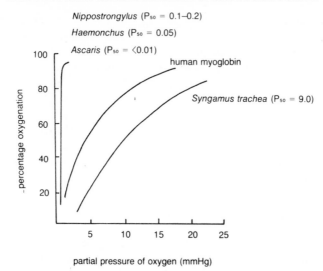

Figure 3.6 Oxygen affinities of some nematode haemoglobins. The values for P_{50} are the pressures of oxygen required to give 50% saturation of the haemoglobin. The figures are for *Ascaris* bodywall Hb and *Haemonchus contortus, Nippostrongylus brasiliensis* and *Syngamus trachea* perienteric (pseudocoelomic) fluid Hb.
Redrawn from Atkinson, H. J. (1976) in *The Organisation of Nematodes.* (editor N. A. Croll), Academic Press, 243–272.

animal-parasitic nematodes have revealed that the presence of these respiratory pigments probably allows the parasite to inhabit regions of low oxygen tension and yet still obtain oxygen for use in certain tissues rather than for the whole animal. *Ascaris* contains two distinct types of haemoglobin; one form of the pigment is found in the perienteric fluid and serves as a source of haematin, while a second form is found in the body wall and is probably truly respiratory in function. Characteristically, the respiratory pigments of many nematode parasites (*Ascaris, Haemonchus* and *Nippostrongylus*) possess a high affinity for oxygen (figure 3.6). By contrast some species, such as *Syngamus trachea* and perhaps *Camallanus trispinosus*, contain haemoglobin with a low affinity for oxygen, and this is thought to be related to the relatively high oxygen tensions of the environments inhabited by these parasites, e.g. the trachea of birds in the case of *Syngamus*.

We may assume that parasites with respiratory pigments that have a high affinity for oxygen possess an adaptive advantage for life in habitats of low oxygen tension, such as the vertebrate alimentary canal. However little information is available on the release of oxygen from such haemoglobins. Partial deoxygenation of some nematode haemoglobins has been demonstrated *in vitro* in conditions which might limit oxygen consumption, but no data are available for the release of oxygen *in vivo*. Studies on the pharyngeal haemoglobin of a free-living marine nematode, *Enoplus brevis*, indicate that deoxygenation of haemoglobin will take place both at low oxygen tensions and during periods of activity by the worm—in this way the respiratory pigment could guarantee oxygen release at the tissues in a time of need. Clearly more information must be gathered on the respiratory pigments of parasitic animals before any substantial generalizations can be made.

Pentose-phosphate pathway in parasites

This pathway is of relatively minor significance in carbohydrate catabolism. Hexoses are converted via a series of reactions to pentoses, with a net yield of two moles of NADPH, and hence six moles of ATP, per mole of glucose metabolized. The biosynthetic nature of the pathway is probably more important, particularly with regard to nucleic acid synthesis (figure 3.7).

Parasites that possess a functional pentose phosphate pathway include *Trypanosoma cruzi*, *Plasmodium berghei*, *Echinococcus granulosus* and *Ascaris lumbricoides*. Unfortunately there has been little attempt to verify

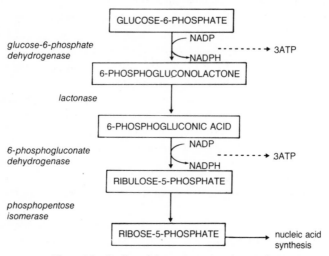

Figure 3.7   Outline of the pentose phosphate pathway.

the presence of the pathway in other parasites and its significance to the parasitic mode of life remains obscure.

The glyoxylate pathway

Many micro-organisms and higher plants use the gloxylate cycle for carbohydrate synthesis from two-carbon precursors, notably from lipid via acetyl-CoA (figure 3.8). The key enzymes of the pathway, isocitrate lyase and malate synthetase, are not found in higher animals. As far as parasites are concerned, this cycle is known to operate in only two species, *Fasciola hepatica* (adult) and *Ascaris lumbricoides* (eggs), though complete evidence for a functional cycle in the former species is lacking.

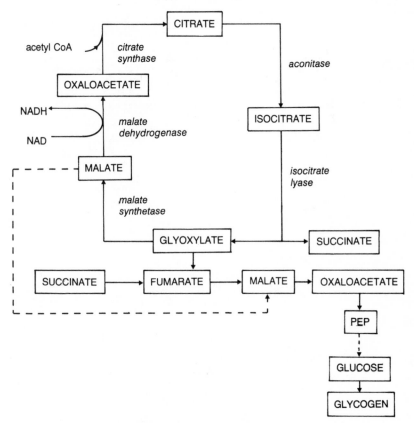

Figure 3.8   The glyoxylate cycle.

SUMMARY

1. Glycogen forms the most common stored polysaccharide in parasites, though some protozoans store amylopectin. Simple sugars are found in parasite tissues in low concentrations only.

2. Glycolysis (the anaerobic catabolism of carbohydrate to yield ATP) is a key pathway for energy metabolism in a great many parasites, particularly those that inhabit regions of reduced oxygen tension. Parasites liberate a wide variety of end-products from glycolysis.

3. Two pathways of CO_2 fixation have been described in parasites of the vertebrate gut. This process involves either "malic enzyme" or phosphoenolpyruvate carboxylase.

4. The tricarboxylic acid cycle is present in some parasites, but incomplete or absent from others. Some helminths use part of the cycle in the reverse direction, reducing malate to fumarate and then to succinate, which is then excreted.

5. Despite the importance of anaerobiosis, many parasites take up oxygen when it is available. The metabolic importance of oxygen is a matter of some controversy.

6. Functional cytochrome systems have been described in a number of parasites. A common feature of these systems is branching of the chain to give more than one terminal oxidase. The significance of cytochromes to anaerobic parasites is not clear.

7. The occurrence of minor pathways of carbohydrate metabolism in parasites has been studied infrequently.

CHAPTER FOUR

PROTEINS, LIPIDS AND NUCLEIC ACIDS

Introduction

Although a detailed account of the biochemistry of proteins, lipids and nucleic acids is outside the scope of this book, it is nevertheless considered essential that the student of parasitology should be aware of the general direction of much of modern research into parasite organisation at the macromolecular level. What must also be made clear is that the biochemistry of these fundamental macromolecules is poorly understood when we compare the information available for parasites with that for bacteria and mammals.

This chapter provides a brief insight into the major features of macromolecular biochemistry of parasites. One of the basic interests in parasite biochemistry is the quest for outstanding differences in the essential biochemical pathways of parasites and their hosts. Such differences could then be exploited chemotherapeutically and serve to eradicate or minimize the effects of pathogenic parasites of man and his domestic animals.

Proteins

Protein composition

Proteins are composed of long chains of L-α-amino acids, each linked by a peptide bond (figure 4.1). Only twenty of the known amino acids commonly occur in proteins, nevertheless protein structure shows considerable variation, both in amino acid sequence and in gross morphology. Proteins are involved in a number of fundamental functions in all tissues, i.e. in a structural and support role, in the formation of contractile elements, and as enzymes, hormones, antigens and antibodies. Some proteins, such as the glycoproteins and the lipoproteins, contain

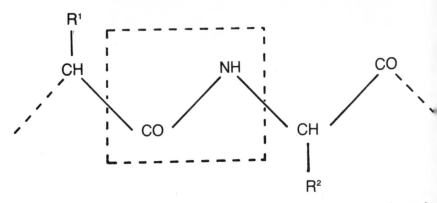

Figure 4.1   The peptide bond linking adjacent amino acids in a protein chain. R^1 and R^2 will be different amino acid residues, such as an aliphatic or an aromatic group, depending upon the particular amino acid.

carbohydrates and lipids respectively, whereas others may have a metallic prosthetic group, e.g. iron in haemoglobin.

The total protein content of parasite tissues varies from around 20% of the dry weight to as much as 80%. Protein content is derived either from the total nitrogen content, multiplied by a factor of 6.25 (accuracy will depend upon the amounts of non-protein nitrogen present), or by direct chemical determination using the Lowrie colourimetric method.

Tissue proteins fall into two broad categories—soluble and insoluble, or particulate. The particulate proteins are associated with the cell membranes or with intracellular membranous structures. The soluble proteins include enzymes, hormones and antigens (parasites do not have antibodies). The bulk of the particulate proteins are involved in a structural and support role and include collagen, keratin and the scleroproteins. Collagen is a common constituent of parasite tissues, occurring particularly in the nematode cuticle and egg shell, where it confers a high degree of mechanical strength and impermeability. Chemical analysis of *Ascaris* cuticular collagen shows it to differ from vertebrate collagens in having only small quantities of hydroxyproline and hydroxylysine; *Ascaris* muscle collagen, on the other hand, more resembles the vertebrate forms in its chemical composition. Keratin itself is not found in parasites but keratin-like proteins do occur in the tissues of some nematodes and acanthocephalans, as well as in the hooks of some cestode scolices. Scleroproteins are constituents of the walls of some parasitic protozoan spores but are more common as the quinone-

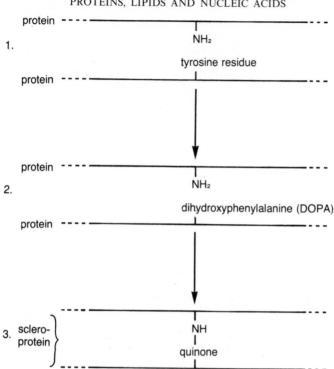

Figure 4.2 Simplified scheme of quinone tanning to form protective scleroproteins. Two adjacent protein chains are cross-linked when the O-quinone of one chain reacts with the free amino (–NH$_2$) residues on the other protein chain.

tanned form, sclerotin, found in the egg shells of digeneans and pseudophyllidean cestodes and in the metacercarial cyst walls of some digeneans. Quinone tanning is brought about by the action of a phenolase which converts an o-phenol to the equivalent o-quinone in the presence of oxygen (figure 4.2). The o-quinone then cross links with an adjacent protein by reacting with free NH$_2$ groups to produce stable sclerotin, which is typically brown in colour. The precise mechanisms involved in quinone tanning are not understood, but it is widely used in the helminths as a means of protecting the otherwise fragile eggs.

Amino acid content

The amino acid content of parasite proteins and of the free pool of unbound acids shows no major departure from that in non-parasitic

Table 4.1 Amino acid biosynthesis in selected protozoan parasites.

Organism	Source molecule	Amino acids synthesized
Trypanosoma rhodesiense	carbohydrate	alanine, glycine, serine, aspartic acid, glutamic acid.
T. cruzi	serine	alanine, aspartic acid, glutamic acid, glycine, cysteine, threonine.
T. gambiense	glucose	alanine.
	alanine	cysteine, taurine.
	aspartate	serine.
	glutamic acid	cysteine, taurine, threonine.
	arginine	taurine, glutamic acid, proline.
T. brucei	proline	alanine, aspartic acid.
Crithidia fasciculata	methionine, cysteine	threonine.
Leishmania tarentolae	arginine	proline.
L. donovani	proline	arginine.
Plasmodium knowlesi	carbohydrate	alanine, aspartic acid, glutamic acid.
	methionine	cysteine.
	serine	methionine.
Entamoeba histolytica	carbohydrate	alanine, aspartic acid, glutamic acid.

animals. The requirement for amino acids by parasites has not been thoroughly examined, as a consequence of which, it is not possible to list the essential amino acids for any parasite species. Probably the greatest body of information on requirement for, and biosynthesis of amino acids concerns the parasitic Protozoa, particularly the trypanosomes and the malarial parasites. Despite this, even for the Protozoa, there is only limited information available at present (table 4.1). For instance, *Leishmania tarentolae* cannot synthesize arginine from proline, the former therefore being an essential requirement. On the other hand, *L. donovani* does carry out this synthesis. *Crithidia oncopelti* is unusual in that it will grow in culture with methionine as its only source of amino acid, but this property is thought to be due to the presence of the "bipolar body" in this parasite, which recent evidence suggests is probably an endosymbiotic bacterium. Other trypanosomatid parasites require much more complex mixtures of amino acids for growth in culture.

It has proved difficult to cultivate metazoan parasites *in vitro* and therefore our knowledge of essential amino acids and biosynthetic pathways remains scanty. The cestode *Hymenolepis diminuta* has been cultured from cysticercoid to adult in a highly complex medium, containing many amino acids. However, subsequent attempts to repeat Berntzen's (1961) culture procedures with this tapeworm have met with no success. Culture media that have been used to grow other species of helminth parasite usually contain crude preparations such as horse serum, yeast or liver extract, or protein hydrolysates, Each of these is provided with a full complement of amino acids and therefore provide no clue as to which amino acids are required and which can be synthesized from precursors.

In many parasites, transamination reactions, involving aminotransferases, are common systems for the interconversion of amino acids, e.g. α-ketoglutarate to glutamic acid (figure 4.3) and pyruvate to alanine. For both these reactions a wide range of amino acids can act as amine donors. The α-ketoglutarate system operates in *Trypanosoma cruzi*, *L. donovani*, *Entodinium shaudini*, *Schistosoma japonicum*, *Fasciola hepatica*, *Hymenolepis citelli*, *H. diminuta*, *H. nana* and *Ascaris lumbricoides*. Other keto acids can also act as transamination substrates, e.g. oxaloacetate is involved in amino acid biosynthesis in *H. diminuta*, *F. hepatica*, *S. japonicum* and *A. lumbricoides*.

1. General reaction:

$$\text{amino acid}_1 + \text{keto acid}_1 \rightleftharpoons \text{amino acid}_2 + \text{keto acid}_2$$

2. α-ketoglutarate: glutamic acid transamination:

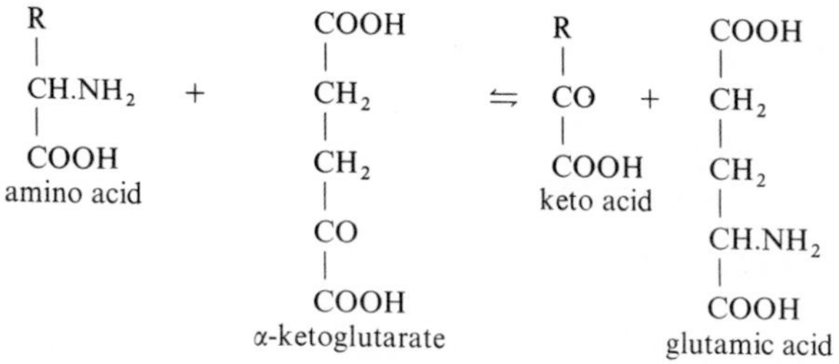

Figure 4.3 Transamination schemes. (1) General reaction. (2) α-ketoglutarate: glutamic acid transamination.

Protein synthesis

Studies on the synthesis of proteins by parasites have proceeded on two fronts, one involving the study of incorporation of radioactively-labelled precursors into the protein of intact organisms, the other utilizing cell-free ribosomal preparations. Once again we find that the majority of studies have utilized the parasitic Protozoa, especially the trypanosomes. These haemoflagellates, like mammals, possess both mitochondrial and cytoplasmic protein synthesis. The mitochondrial system, like that of bacteria, is sensitive to the potent inhibitor chloramphenicol, whereas the cytoplasmic system is inhibited by cycloheximide. This state of affairs holds true for the majority of protozoans studied, with the possible exception of mitochondrial protein synthesis in *Crithidia fasciculata*.

Ribosomal preparations of *Crithidia oncopelti*, containing 86S ribosomes, can activate a number of amino acids at pH 6.8 in the presence of Mg^{++}. The cytoplasmic ribosomes of *C. fasciculata*, ranging in size from 83 to 88S, also depend upon Mg^{++} for amino acid activation. Additional studies on the malarial parasites, *Plasmodium berghei* and *P. knowlesi*, show that there is considerable similarity between their cytoplasmic protein synthesis and that of mammals.

Virtually nothing is known of the details of protein synthesis in helminth parasites. This should, however, prove to be an interesting area of research since many helminths must have a remarkably high rate of protein synthesis associated with their prolific level of egg production, e.g. the female schistosome sheds 10% of her body weight every twenty-four hours in egg production alone. Many helminths are known to absorb and incorporate a wide variety of substrates into their tissue proteins, but in general such an approach tells us little about the mechanisms involved. There is, however, no reason to believe that the fundamental mechanisms of protein synthesis differ in parasitic and non-parasitic animals.

Protein and amino acid metabolism

Although the majority of internal parasites rely heavily upon carbohydrate metabolism as their basic source of energy, a few examples are known in which protein and amino acid metabolism achieve some degree of importance, particularly under emergency conditions. Incubation *in vitro*, in the absence of carbohydrate, leads to an increased production of ammonia in a number of parasites (*Trypanosoma cruzi*, *Trichomonas vaginalis*, and *Plasmodium gallinaceum*) suggesting active protein metabolism. Starved *Entodinium caudatum* may also use protein metabolism for energy production and *Leishmania tarentolae* shows an increased oxygen uptake in the presence of some amino acids. This latter organism is unusual in its requirement for the imino acid, proline, as a major energy source, oxidising it first to glutamate and finally to alanine and CO_2. Other protozoan parasites, including *Trypanosoma rhodesiense* and *Plasmodium lophurae*, can also utilize amino acids for energy metabolism and there is a suggestion that some nematode larvae metabolize proteins under emergency conditions.

The major end-products of nitrogen metabolism will be considered in chapter 5. Suffice it to say here that the majority of protozoans excrete ammonia while metazoan parasites excrete ammonia and urea to a greater or lesser extent. Many parasites excrete amino acids, but the mechanisms involved and the significance of this remain obscure.

Respiratory proteins

A wide variety of parasites contain haemoglobin. In the platyhelminths, haemoglobin occurs within the tissues. In the nematodes, it may also be found in the pseudocoelomic (perienteric) fluid—a situation reminiscent of the annelids. Parasite haemoglobins, particularly those of the nematodes, are characterized by their high affinity for oxygen, suggesting a true respiratory function. Despite this, it has been observed that the body-fluid haemoglobin of some nematodes is curiously reluctant to release its bound oxygen, and this has given rise to considerable speculation as to the real function of this family of proteins. Alternative functions have been postulated and they include: peroxide elimination via methaemoglobin; the formation of a pool from which other haem compounds might be synthesized; and photosensitivity (as in *Mermis subnigrescens*). Other authors have concluded that these parasite haemoglobins are, in fact, vestigial and without function. Clearly this is an interesting area of research which might benefit from continued attention.

The cytochromes are respiratory pigments (chapter 3) which contain a haem prosthetic group. The structure of cytochromes has been examined in only a few parasites, including the trypanosomes and a small number of helminths (*Moniezia* and *Ascaris*). Spectral and amino acid analyses reveal certain differences between parasite and vertebrate cytochromes, e.g. the amino acid, trimethyl-lysine, a common constituent of trypanosome cytochromes, is absent from vertebrates.

Table 4.2 Classification of naturally occurring lipids.

Fatty acids
      saturated
      unsaturated
      branched chain

Glycerol derivatives
      glycerides (mono, di and tri)
      phosphatides
      plasmalogens

                                phospholipids

Sphingosine derivatives
      sphingomyelins
      ceramides
      cerebrosides
      gangliosides

Sterols (steroidal alcohols)

Terpenes

Waxes (long chain alcohols esterified to fatty acids)

Conjugated lipids
      lipoproteins
      lipopolysaccharides

Intraerythrocytic stages of the malarial parasites produce a haemoprotein-containing pigment within the parasitized red cell. The structure and function of this unique pigment is at present unknown.

Lipids

Lipid composition

Lipids are a highly diverse and heterogeneous group of compounds, unified only by their common extraction from animal tissues by the use of organic solvents, such as methanol, chloroform and acetone. A simplified classification of naturally occurring lipids is given in table 4.2.

Fatty acids Saturated (with no double bonds) and unsaturated (with varying numbers of double bonds) fatty acids occur in glycerides and phospholipids, and small quantities of free fatty acids can be extracted from all animal tissues. Some nematodes may possess unusually large pools of these free fatty acids e.g. *Nippostrongylus brasiliensis*, 60% of total lipid; *Strongyloides ratti*, 67% of total lipid. More usually, the free fatty acid pool represents less than 10% of the total lipid content, and this will often include glycerides as well. In general, parasite fatty acids are no different from those found in non-parasitic animals.

Glycerides In mammals, glycerides—composed of a glycerol back-bone and either one, two or three fatty acids—form the major storage element of lipid. In parasites, stored lipid is rather less abundant than in other animals. The parasitic Protozoa possess only small amounts of triglyceride, though it is interesting to note that the culture forms of some trypanosomes contain consistently more triglyceride than the equivalent bloodstream forms; the reason for this is not clear. Triglycerides may act as an energy store in the encysted stages of some protozoans, such as the oocysts of *Eimeria*, whose metabolism is probably aerobic. The glyceride content of helminth parasites has not been adequately investigated for more than a small number of species. Approximately 30% of schistosome lipids are composed of glycerides and free fatty acids, while somewhat higher figures have been obtained for some nematodes, cestodes and acanthocephalans.

Phospholipids Phospholipids are the major lipid components of cell membranes and are derivatives of phosphatidic acid (figure 4.4). In the phospholipids, a variety of polar groups are found, e.g. choline in phosphatidylcholine, ethanolamine in phosphatidylethanolamine, serine in phosphatidylserine, inositol in phosphatidylinositol and glycerol in phosphatidylglycerol and diphosphatidylglycerol or cardiolipin. It is likely that most parasites contain phospholipid in their membranes but only a few species have been examined in any detail. The most commonly occurring phospholipids are phosphatidylcholine, phosphatidylethanolamine, phosphatidylserine, phosphatidylinositol, cardiolipin, sphingolipids and plasmalogens.

Sterols The most abundant sterol in parasite tissues is cholesterol—it is the predominant sterol in trypanosomes, *Entamoeba*, *Plasmodium* and *Trichomonas*. An alternative sterol, ergosterol, is found in place of cholesterol in the culture forms of some of the trypanosomes (*T. cruzi*, *T. lewisi* and *T. rhodesiense*). The reason for this switch in sterols during the life-cycle is not known. All the helminths that have been examined possess cholesterol, as well as a variety of cholesterol esters, whose distribution is somewhat uneven. It appears that sterols in general are

(a)

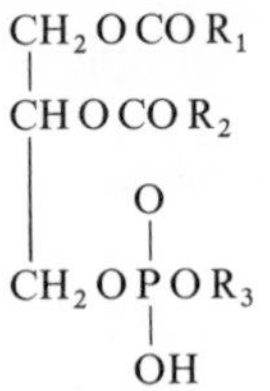

$$CH_2 O CO R_1$$

$$CH O CO R_2$$

$$CH_2 O \overset{O}{\underset{OH}{P}} O R_3$$

(b)

Phosphatidic acid: $R_3 = H$; R_1 and R_2 are long aliphatic chains.
Phosphatidylglycerol: $R_3 = CH_2CHOHCH_2OH$
Cardiolipin(diphosphatidylglycerol):

$$R_3 = CH_2 CHOH CH_2 O \overset{O^-}{\underset{O}{P}} O CH_2 - \overset{CH_2 OOC R_1}{\underset{\underset{O}{|}}{\overset{|}{CH OOC R_2}}}$$

$$\underset{COO^-}{}$$

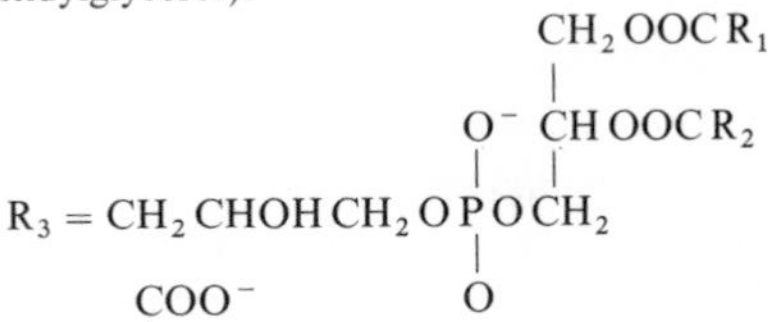

Phosphatidylserine: $R_3 = CH_2 CH NH_3^+$
Phosphatidylethanolamine: $R_3 = CH_2 CH_2 NH_3^+$
Phosphatidylcholine: $R_3 = CH_2 CH_2 N_+ (CH_3)_3$

Figure 4.4 (a) The general structure of phospholipids. In the parent molecule, phosphatidic acid, $R_3 = H$. (b) Some common examples.

not synthesized by parasites and are therefore essential dietary requirements.

A small number of parasites contain pigmented carotenoid material, which is derived from cholesterol. This is not uncommon in the Acanthocephala, e.g. the cystacanth larvae of *Polymorphus minutus* are characteristically surrounded by a bright orange capsule which is rich in carotenoids. It is often speculated that this pigmentation makes the parasite more obvious to the final host, a duck, when it is feeding on fresh water shrimps (the intermediate hosts for the parasite). A few digeneans and nematodes also contain carotenoids.

Waxes Wax esters have been extracted from the tissues of only a small number of parasites, such as acanthocephalan larval stages. There is some doubt whether waxes occur in the parasitic Protozoa.

Lipid biosynthesis

Parasites in general appear to possess only a limited ability to synthesize long chain fatty acids *de novo* and therefore rely to a great extent upon the host to provide their essential requirement. Mammals can synthesize long chain fatty acids by a reversal of the β-oxidation process, using acetyl-CoA as the basic two-carbon foundation. Although this *de novo* pathway may occur in some trypanosomes, the majority of parasites (Protozoa and helminths alike) are restricted to conservative chain lengthening of absorbed fatty acids by the addition of acetate units. Chain elongation is a feature of lipid biosynthesis in a number of parasites including *Leishmania*, some trypanosomes, trichomonads, *Crithidia*, *Plasmodium*, *Schistosoma* and *Hymenolepis*. The nematode, *Ascaris lumbricoides*, possesses a limited facility for lipid synthesis using malonyl-CoA. As a consequence of the overall inability of parasites to synthesize their long-chain fatty acids, it is not surprising to find that the fatty acid content of a parasite closely resembles that of the host in which it resides. This is exemplified by *Hymenolepis diminuta*, whose fatty acid and glyceride content may be altered by feeding the host experimental diets with either an excess or a deficiency of various lipids. Both palmitic acid ($C_{16:0}$) and linoleic acid ($C_{18:2}$) proved to be highly variable under such experimental conditions.

Schistosoma mansoni and Fasciola hepatica are both capable of chain elongation by the addition of acetate units to absorbed fatty acids. Comparison of rat liver fatty acid content with that of *Fasciola* grown in rats shows considerable similarity, with the exception of the occurrence of two acids, eicosenoic ($C_{20:1}$) and eicosadienoic ($C_{20:2}$), which appear in the worm pool but are absent from the host. *Fasciola* has a high rate of phospholipid synthesis from the precursors phosphate and glycerol, which are readily incorporated into lysolecithin, phosphatidylcholine, phosphatidylinositol, phosphatidylserine, phosphatidylethanolamine, ethanolamine plasmalogen and cardiolipin.

Threonine is an important precursor for acetate and acetyl-CoA in *Trypanosoma* and *Crithidia*. The postulated biosynthetic pathway for these compounds involves the intermediate, α-amino-β-oxybutyrate, and is inhibited by tetraethylthiouram, a powerful growth inhibitor of some trypanosomes.

There is little information on the mechanisms employed by parasites for the introduction of double bonds (desaturation) into long-chain saturated fatty acids, nor is much known of the processes involved in triglyceride synthesis. *Hymenolepis* rapidly incorporates radioactively labelled glycerol, palmitic acid and oleic acid ($C_{18:0}$) into triglycerides and phospholipids. This synthesis is thought to occur via steps involving α-glycerophosphate, phosphatidic acid and diglyceride. Additionally, studies on *Fasciola* have demonstrated the rapid incorporation of serine,

Table 4.3 The structural components of the nucleic acids, DNA and RNA.

	Synthetic sequence	Components of DNA	Components of RNA
bases	*de novo* synthesis, salvage, or dietary requirement	*purines* adenine guanine *pyrimidines* cytosine thymine	*purines* adenine guanine *pyrimidines* cytosine uracil
nucleosides	base + sugar	*deoxyribonucleosides* (sugar = deoxyribose) deoxyadenosine deoxyguanosine deoxycytidine deoxythymidine	*ribonucleosides* (sugar = ribose) adenosine guanosine cytidine uridine
nucleotides	nucleoside + phosphate	*deoxyribonucleotides* deoxyadenosine-monophosphate deoxyguanosine-monophosphate deoxycytidine-monophosphate deoxythymidine-monophosphate	*ribonucleotides* adenosine-monophosphate guanosine-monophosphate cytidine-monophosphate uridine-monophosphate
nucleic acids	polymerization of nucleotides	DNA	RNA

glucose and oleic acid into phospholipids and glycolipids, including cerebrosides. Phosphatidylethanolamine syntheses in *Hymenolepis diminuta* proceeds via the decarboxylation of phosphatidylserine. Rapid phospholipid synthesis is a characteristic feature of many protozoans.

The ability of parasites to synthesize sterols is apparently limited, and most rely on their hosts to provide these essential compounds. Exceptions to this include *Crithidia fasciculata* and some trypanosomes.

Lipid catabolism

Mammals catabolise lipids aerobically by β-oxidations but, since many parasites are facultative or even obligate anaerobes, we might justifiably expect limited use of such oxidative metabolism. This expectation is borne out, and very few parasites make use of lipid for energy metabolism. Those that do include *Trypanosoma lewisi*, *Leishmania brasiliensis*, *Litomosoides carinii*, *Echinococcus granulosus*, *Taenia hydatigena*, *Ascaris* and the free-living stages of some larval nematodes. Other parasites, like *Hymenolepis*, do not possess the full complement of β-oxidation enzymes. Anaerobic catabolism of fatty acids has been recorded in some of the rumen-dwelling ciliate Protozoa, but this is most unusual.

Nucleic acids

Nucleic acid composition

The two ubiquitous nucleic acids in all animal tissues are deoxyribonucleic acid (DNA) and ribonucleic acid (RNA). Most of the DNA is found within the cell nucleus and it constitutes the basic genetic information or genome. DNA is a double helical molecule with specific pairing of bases between each strand of the double helix so that guanine and cytosine are always linked, as are adenine and thymine. It is the sequence of base pairs in the DNA of any organism that represents foundation for its heritable characteristics. Small quantities of DNA may also be found outside the nucleus, often associated with mitochondria (mtDNA). Mitochondrial DNA probably codes for ribosomal RNA synthesis. RNA occurs in three basic forms, messenger RNA (mRNA), transfer RNA (tRNA) and ribosomal RNA (rRNA). Structurally, RNA is similar to DNA but contains ribose in place of deoxyribose, uracil in place of the closely related pyrimidine, thymine, and is frequently single rather than double stranded (table 4.3). The sequence of base pairs in nuclear DNA determines the base sequences of RNA (transcription), and the latter forms the template for protein synthesis (translation). In this sequence of events mRNA is involved in transcription and tRNA in translation. The final step, protein synthesis, is ribosomal in location.

The field of molecular biology, with its core in the biochemistry of the nucleic acids, has developed at a truly dramatic rate and parasitic organisms have not been overlooked in this modern approach to biology. However, the trypanosomatid flagellates have received considerably more attention than any other group of parasites. This is, in part, easily attributable to the presence in these protozoans of an unusual form of extranuclear DNA, known as kinetoplast DNA (kDNA).

The cell nuclei of all parasites are presumed to contain DNA— considerable evidence for this comes from Feulgen staining and, more recently, from direct chemical determinations. There is no reason to believe that parasite DNA contains any unusual bases—determination of the base composition, using a range of techniques, including buoyant density determination in caesium chloride gradients in the analytical ultracentrifuge, melting point determination and direct chemical analysis, reveals the expected range of purines and pyrimidines. However, there is some doubt as to whether any of these methods would expose the presence of any minor components, such as methylated bases.

The base composition of DNA is usually expressed as the percentage of guanine and cytosine relative to the total base content, i.e. the $G+C$ content. Figures for the $G+C$ content of parasite nuclear DNA range from 18 to 61%, while the $G+C$ content for extra nuclear DNA is more restricted, lying between 18 and 41%. These determinations have been made on less than 40 parasite species so that any generalisations are, perhaps, inopportune.

Little is known about the RNA of parasites and studies are restricted almost exclusively to the parasitic protozoans and ascarid nematodes. RNA has been characterized from the chromatoid bodies (composed of symmetrically arranged ribosomes, made up of 30S and 50S subunits) of *Entamoeba invadens*, where it is associated with a protein component. Ribosomal RNA and transfer RNA have been isolated from a small number of protozoans, but very little is known about messenger RNA.

Nucleic acid synthesis

Purine and pyrimidine synthesis Most of the available information on the synthesis of nucleic acids comes from the protozoans and rather little from the helminths. Mammals synthesize the purine ring either *de novo* or from salvage pathways, involving preformed bases or nucleosides. Many parasitic protozoans differ from mammals in that they rely almost exclusively on a dietary source or salvage pathway, since they are apparently incapable of *de novo* synthesis of the purine ring. At least two species, *Crithidia oncopelti* and *Trypanosoma lewisi*, are exceptions to this generalisation and synthesize purines from glycine and serine. Purine salvage pathways have been investigated in *Plasmodium* and *Trypanosoma* and it is apparent that hypoxanthine, adenosine and adenine are freely converted to purines.

In direct contrast to the purines, pyrimidines are synthesized *de novo* by many parasites. All of the species studied so far, most of which are protozoans, synthesize pyrimidines from precursors such as aspartate, bicarbonate and glutamine. Additionally, salvage pathways are important sources of these bases in many of the Protozoa. The enzymes involved in the *de novo* synthesis of pyrimidines by *Plasmodium* have been identified and include orotidine-5-monophosphate pyrophosphorylase, deoxythymidilate synthetase and dihydrofolate reductase.

It is not certain whether helminth parasites can synthesize either purine or pyrimidine bases *de novo* or whether they simply rely on the host to be their source of supply. Several species of cestodes, digeneans and acanthocephalans are freely permeable to a variety of bases, but the mechanisms of acquisition have been examined in one species (*Hymenolepis diminuta*) only (see chapter 2). The difficulties in maintaining helminths in defined culture media for any length of time are undoubtedly responsible for the basic lack of knowledge. All of the data on purine and pyrimidine synthesis in the Protozoa derive exclusively from organisms grown in defined or partially defined media containing radioactively labelled precursor molecules.

Nucleic acid synthesis Synthesis of nucleic acids involves a series of enzymes termed polymerases and virtually nothing is known of these polymerases in parasitic tissues.

Nucleic acid catabolism

Nucleases, such as DNA-ase and RNA-ase are responsible for catabolising nucleic acids to nucleotides, and phosphorylases, phosphatases and hydrolases catabolise these nucleotides to nucleosides and free bases. A variety of these enzymes have been isolated from parasites, including *Crithidia oncopelti*, *Plasmodium*, *Trypanosoma*, *Entamoeba*, *Ascaris* and

Trichuris. The tissues of *Hymenolepis diminuta* contain the β-amino acids, β-alanine and β-aminoisobutyric acid, which are thought to originate from uracil and thymine. Uric acid, a breakdown product of purines, is excreted by a number of helminths, implying active nucleic acid catabolism.

The extranuclear DNA of trypanosomes

Extranuclear or mitochondrial DNA is a common feature of most organisms, and that of parasites shows no fundamental difference from that of free-living animals. The kinetoplast DNA of trypanosomes was the first extranuclear DNA to be described for eukaryotes.

In caesium chloride gradients, kDNA characteristically appears as a rapidly-banding peak which can readily be differentiated from the nuclear DNA (figure 4.5). Most extranuclear DNA is circular in configuration and kDNA is no exception. Kinetoplast DNA can be isolated from the trypanosome mitochondrion intact and visualised under the electron microscope, where its circular nature is clearly

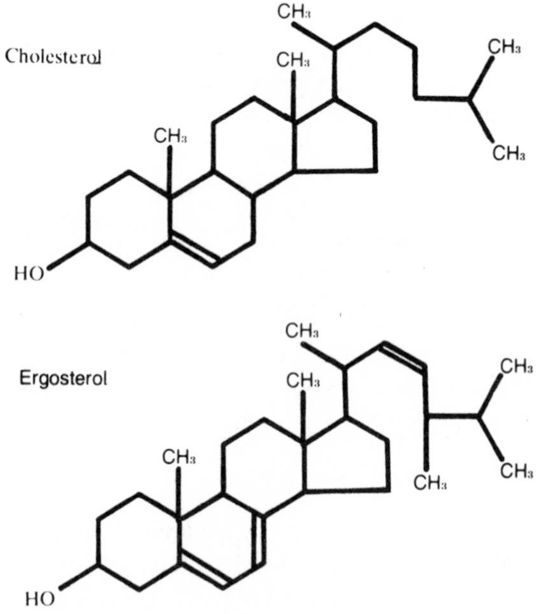

Figure 4.5　The sterols, cholesterol and ergosterol.

A. GROSS MORPHOLOGY OF A GENERALIZED TRYPANOSOME

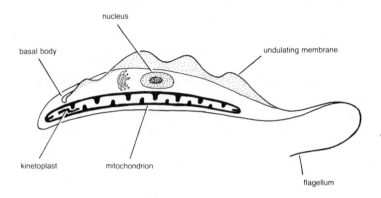

B. MICRODENSITOMETER TRACING OF DNA FROM *TRYPANOSOMA BRUCEI*

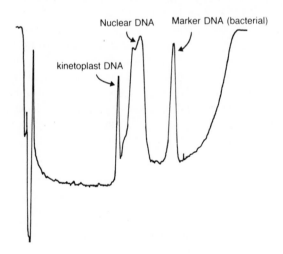

Figure 4.6 The kinetoplast and kinetoplast DNA of trypanosomes. Kinetoplast DNA appears as a rapidly-banding peak distinct from nuclear DNA; marker DNA is an added internal standard of bacterial or mammalian origin. The tracing in B. is derived from the analytical ultracentrifuge and is recorded under U.V. light (260 nm) at equilibration in a caesium chloride gradient.

revealed. In fact, kDNA is composed of two distinct circular elements, referred to as mini-circles and maxi-circles. The diameter of the mini-circles ranges from 0.3—0.8 μm while the maxi-circles have a mean diameter of 11 μm. The smallest mini-circles are typical of the mammalian trypanosomes, while the larger mini-circles occur in the insect-parasitic species. The application of restriction endonucleases (enzymes which cleave nucleic acids in specific places) reveals that there is some degree of nucleotide heterogeneity in terms of sequence in the mini-circular kDNA but the maxi-circles are made up of non-repeating nucleotide sequences. On visualisation with the electron microscope, kDNA is seen as a network of circular components, of which less than 10% is the maxi-circular element; maxi-circles can be completely removed by enzyme treatment without loss of the integrity of the DNA network. In *Crithidia* there are approximately 14 000 mini-circles and 50 maxi-circles in kDNA. The elements of the kDNA network are linked by covalent bonds and the mini-circles are thought to be catenated or fused together, possibly into long chains.

Although the precise function of kDNA is far from clear, it is thought that maxi-circles represent true mitochondrial DNA and therefore code for ribosomal RNA. The function of the mini-circular component of kDNA is even less clear, but it may have a supportive or structural role. Dyskinetoplastic trypanosomes, occurring either naturally or produced by laboratory manipulation, tend to lose their kDNA and, with it, the ability to synthesize mitochondria. Some dyskinetoplastic forms, however, like *T. equiperdum* and *T. evansi*, seem to possess normal kDNA. There is much to learn from this area of research, particularly concerning the role of kDNA in parasitism by trypanosomes.

SUMMARY

1.   The support proteins, collagen, keratin-like proteins and scleroproteins, are present in many parasite tissues; quinone-tanned scleroproteins form the resistant coverings of helminth eggs and cysts. The amino acid content of proteins, the composition of the free amino acid pool, the essential amino acids, amino acid biosynthesis and protein synthesis have been investigated in only a small number of parasites. A few protozoans derive energy from amino acid metabolism but the helminths do not. The function of respiratory proteins, such as haemoglobin, is not known for parasites.

2. The pool of free fatty acids and glycerides is small in most parasites. A variety of phospholipids are present in parasite cell membranes. The sterols, cholesterol and ergosterol, are probably widespread, but have only been isolated from a few protozoans. Parasites generally do not synthesize long-chain fatty acids but rely on a dietary supply of these molecules, which they can modestly chain-lengthen. There is relatively little information on lipid biosynthesis in parasites. Catabolism of lipid, by β-oxidations, occurs in some parasites, but those that inhabit an environment with a low oxygen content lack these pathways.

3. DNA and RNA are normal constituents of parasite tissues. The base composition of DNA has been determined for less than forty parasite species. Extranuclear DNA (kinetoplast DNA) has been extensively studied in the trypanosomes but its function is not understood. *De novo* purine synthesis is lacking in most parasites studied—salvage pathways or a dietary source provide parasites with the essential purines for nucleic acid synthesis. Many parasites synthesize pyrimidines *de novo*. Little is known about DNA or RNA polymerases in parasites.

CHAPTER FIVE

EXCRETORY SYSTEMS, NITROGEN EXCRETION, WATER AND IONIC REGULATION

Introduction

The excretory system of all animals is responsible for maintaining a relatively constant internal environment in the face of an ever-changing external environment. It follows, therefore, that a parasite inside its host's tissues will be inhabiting an environment that is itself rigorously controlled by the excretory and osmoregulatory systems of the host animal. Parasites disperse themselves from host to host and, during this transmission, they will be exposed to a variety of physico-chemically different environmental conditions. Many parasites have a direct life-cycle with one host and a single transmission stage. At the other end of the spectrum are parasites, e.g. the strigeid digeneans, which, during a single life-cycle, infect three different hosts and have two distinct transmission stages. We would thus expect the excretory system of parasites to be able to remove the waste products of their metabolism and carry out an osmoregulatory role when the prevailing conditions demand.

In general terms, excretory systems perform several major physiological functions: they remove toxic end-products of metabolism; they maintain ionic balance; they control the body water content; and they remove toxic foreign substances. The evidence for these functions occurring in parasites is rather limited and a great deal of important work remains to be done on the excretory and osmoregulatory systems of parasitic animals.

Contractile vacuoles in Protozoa

The organelle concerned with water and ion balance in the Protozoa (and in the Porifera) is the contractile vacuole. It plays a significant role in ion and water regulation in freshwater protozoans, but its functional significance to either marine or parasitic species is questionable—both these groups of protozoans inhabit environments that are isosmotic to their internal fluids.

Amongst the parasitic Protozoa, contractile vacuoles are present in a small number of flagellates and many ciliates but they are not typical features of either the sporozoans or the amoebas. The excretion of nitrogenous waste material is probably not a function of the contractile vacuole. Excess nitrogen, in the form of ammonia or urea, is removed by outward diffusion across the cell membrane.

The protonephridial system of platyhelminths

Protonephridia, comprising a blind-ending system of tubules which open via a nephridiopore, are present in a wide range of invertebrate groups. In the protonephridial system of the platyhelminths, a single "flame cell" is located at the terminus of each excretory tubule. The flame cell derives its name from the large number of flagella that beat rhythmically, resembling the flickering of a flame. The protonephridial systems of

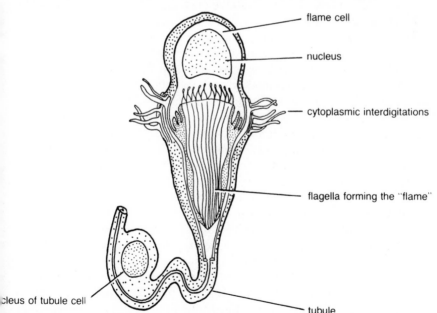

Figure 5.1 The flame cell terminal organ of a digenean. Redrawn from Wilson, R. A. and Webster, L. A. (1974) *Biological Reviews*, **49**, 127–160.

monogeneans, digeneans and cestodes show a fundamental morphological similarity. The ultrastructure of the excretory system has been described for only a small number of platyhelminth species, including *Fasciola hepatica, Himasthla quissetensis, Moniezia expansa, Echinococcus granulosus* and *Diphyllobothrium latum.*

The flagella of flame cells arise from a circle of projections of the cell cytoplasm. These projections interdigitate with similar projections arising from the first cell of the excretory tubule (figure 5.1). The excretory tubule itself is extracellular, the walls of which are formed by a sheet of cytoplasm. There are usually between 50 and 100 flagella in the flame; these are sometimes called cilia, but functional evidence suggests that the term flagella is more appropriate. Each flagellum contains microtubules in a $9+2$ arrangement, and is attached to the cell by a basal body. The flagella beat rhythmically in an undulatory wave motion which is thought to produce the flow of tubule fluid and, perhaps, draws solutes from the surrounding parenchymal tissue into the terminal organ of the excretory system. A nephrostome, consisting of a ring of pores through which solutes might pass, has been described in the tapeworm *Moniezia.* There is no evidence for such a structure in other platyhelminths. Filtration of solutes and suspended material may take place either at the terminal organ itself or in the epithelium of the excretory tubule, a metabolically active region rich in mitochondria, Golgi apparatus and endoplasmic reticulum. The tubules leading from the terminal organ are extremely small and are extracellular, though they were once regarded as intracellular. The tubules are lined with a highly folded cytoplasm in the Digenea and with microvilli in the Cestoda.

The excretory system in the Digenea (figure 5.2) tends to be simple in larval worms and increases in complexity during development, e.g. the miracidium of *Fasciola hepatica* has two terminal organs, each opening to the outside separately, but in the adult liver fluke, there are many hundreds of terminal organs, each with a tubule that enters the larger collecting ducts which then empty into a single excretory bladder. The cestodes are well supplied with terminal organs and excretory tubules and, in addition, there are large ventral and dorsal excretory canals which run the length of the strobila, connected by transverse canals.

Although we refer to the protonephridial system of the platyhelminth parasites as an excretory system, there is remarkably little evidence to support its status as such. Furthermore, it has been debated, at considerable length, whether this system is involved in osmoregulation— we shall examine the evidence for this later. A functional excretory system must carry out selective filtration and reabsorption of solutes in order to regulate the chemical content of the excreta. It has been postulated that filtration in the protonephridial system is effected at the region of interdigitation between the flame cell and the excretory tubule. The flagellar beat may draw solutes into the flame cell and move them along the tubule. It is calculated that a sufficiently large filtration pressure could be developed across the interdigitating membranes for filtration to occur, but there is no direct evidence as yet. There is limited evidence that nitrogenous waste material is transported within the protonephridial system. In *Moniezia expansa*, ammonia is thought to be removed from the body by diffusion through the tissues and across the tegument, while there is positive evidence for the removal of waste

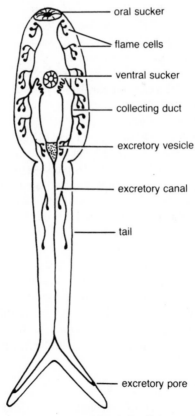

Figure 5.2   The digenean excretory system, illustrated in a fork-tailed cercaria.

nitrogen by the protonephridia of *Hymenolepis diminuta*. The latter tapeworm has been the subject of one of the few analytical studies on the excretory system of parasites. Wilson and Webster developed a technique of micropuncture that allowed them to withdraw small volumes of protonephridial fluid from the longitudinal canals. Chemical analysis of volumes less than $10\,\mu l$ revealed the presence of organic ions (Na^+, K^+, Cl^-, CO_3) equivalent to 1.15% NaCl, amino acids (12 mg/ml), ammonia (0.04 mg/ml), urea (0.5 mg/ml) and lactic acid (0.7 mg/ml). The occurrence of these substances in the excretory canals is consistent with the notion that the protonephridial system of *Hymenolepis* is an excretory system. However, it remains to be established that filtration takes place at the flame cell and that selective reabsorption and secretion (both prerequisites of a true excretory system) occur within the protonephridial ducts. Glucose, lactate and urea are all reabsorbed in the canals of *Hymenolepis*. The flow of fluid within the excretory canals of *Hymenolepis* has been shown to depend upon worm body movements and the excretion rates of the major waste products, in mg/mg dry weight/h, are: lactate 19.4; urea 1.02; and ammonia 0.08. The significance of the protonephridial system in the excretory physiology of cestodes cannot be fully understood until further critical and analytical studies have been made.

The excretory physiology of the Digenea has been inadequately investigated, and we are left unclear as to the precise function of the protonephridial system. Lipid droplets are present in the excretory ducts of *Fasciola hepatica*, *Diplostomum spathaceum*, *Apatemon gracilis*, *Holostephanus luhei* and *Echinostoma revolutum*, whereas no lipid is observed in the protonephridial system of tapeworms. Alkaline phosphatase activity is associated with the excretory system of several digeneans, but this may be involved in the hydrolysis of sugar phosphates in the tubules rather than any direct excretory function. There is no experimental evidence to suggest that nitrogenous or other metabolic waste products are actually excreted by the digenean protonephridial system and it is likely that these substances are lost by diffusion across the tegument. Osmoregulation may be a more important function of the system, particularly in the free-living larval stages.

Excretion in the Acanthocephala

The Acanthocephala are unusual in that the majority possess no discrete excretory organs or system, with the exception of the members of the order Archiacanthocephala, which have a protonephridial system comparable in its structure to that of the platyhelminths. It must be assumed that metabolic waste products are removed by diffusion across the membranes of the cuticular pores, described in chapter 2.

Excretion in the nematoda

The excretory system of the parasitic nematodes (in no way different from free-living forms) is not protonephridial, but consists, in the primitive condition, of gland cells that open to an excretory pore via an excretory vesicle. In the more advanced condition, a canal system, often H-shaped, opens to the outside at a single pore; gland cells are commonly associated with these canals. Experimental evidence points to this system functioning primarily in an osmoregulatory, rather than an excretory role.

End-products of nitrogen metabolism in parasites

The major source of excess metabolic nitrogen derives from proteins, amino acids and, to a lesser extent, from purines and pyrimidines. While the majority of parasites excrete ammonia, and thereby resemble free-living freshwater animals, a variety of other excretory products have been detected, the most common of which is urea.

The majority of Protozoa examined excrete ammonia and we assume that their small size obviates the need to develop a complex nitrogen detoxication system. Urea may be produced in small quantities, but in those species that have been studied, such as *Crithidia fasciculata*, there is no urea cycle (figure 5.3), presumably due to the absence of key enzymes like transcarbamylase. Amino acids also appear in the excreta of many protozoans, evidenced by the occurrence of amino acids in culture media used for growing protozoans *in vitro*. It is not certain that these amino acids are true excretory products and they may simply represent counterflow mechanisms involved in the uptake of nutrient amino acids. In the bloodstream-form of *Trypanosoma gambiense*, absorbed glucose is rapidly converted to alanine, which is presumably excreted. Extensive radiolabelling studies are required to clarify the role of the amino acid pool in excretion.

The platyhelminth parasites excrete a range of organic molecules derived from nitrogen metabolism: ammonia is the major waste product for some species (*Lacistorhynchus tenuis*, *Moniezia benedeni*); ammonia and urea are excreted by *Fasciola hepatica* and *Schistosoma mansoni*;

ammonia and uric acid are excreted by *F. gigantica* and *Paramphistomum explanatum* and the cestodes *Cysticercus tenuicollis*, *Echinococcus granulosus* and *Taenia taeniaeformis* excrete urea and uric acid. *In vitro* many helminths, like the Protozoa, release amino acids into the culture media, but the origins of these amino acids remain to be determined.

Ammonia is produced by parasites by a number of well described pathways which include amino acid oxidases (*Plasmodium berghei, Hymenolepis diminuta, Ascaris lumbricoides*), mono-amine oxidases (*Balantidium coli, Schistosoma mansoni*), glutamate dehydrogenase (*Ascaris, Plasmodium berghei, P. lophurae, Microphallus pygmaeus*), and urease (*Lacistorhynchus tenuis*). Urea formation in vertebrates depends upon the urea cycle (figure 5.3), but its occurrence in parasitic animals is the subject of some controversy. Many parasites contain some of the enzymes of the cycle and some cycle intermediates, but whether the cycle is operable in any parasite is not known. The enzymes arginase and ornithine transcarbamylase are present in the cestode *Hymenolepis*

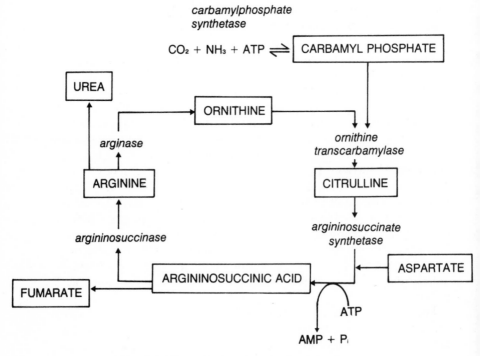

Figure 5.3 The production of urea by the urea cycle.

diminuta, but the worm cannot synthesize the key intermediate carbamyl phosphate. Current opinion is that the functional significance of the urea cycle to parasites is, in general, negligible but that those cycle enzymes that do occur may be involved in alternative biosynthetic activities.

Virtually nothing is known of the excretory products of acanthocephalans. *Moniliformis dubius* contains only one enzyme of the urea cycle, ornithine transcarbamylase, so the cycle probably does not function. Almost all of the parasitic nematodes excrete ammonia, but urea is excreted by some species.

Ionic regulation and water balance

Parasites, unlike free-living animals, are not normally confronted with the major physiological problem of maintaining their water content or the ionic composition of their body fluids at levels very different from the external environment. The greatest osmotic problem faced by parasites will occur during transmission between hosts—transmission stages of parasites may inhabit a marine environment, a freshwater environment or a truly terrestrial environment for varying lengths of time. Inside the host, the parasite inhabits an environment that is approximately isosmotic with its own tissue fluids. Ectoparasites, such as protozoans, almost all of the monogeneans, a single digenean (*Transversotrema patialense*) and the many arthropods, inhabit the external surfaces of aquatic animals. If the host is a freshwater animal the parasite will be faced with a severe water and ion balance problem since the external environment will be hypotonic to its tissue fluids. A similar problem will be shared by the larval stages of platyhelminths, such as eggs, miracidia, oncomiracidia, cercariae and coracidia, which are dispersed in freshwater. The need to resist dessication is a related problem common to protozoans, platyhelminths and nematodes parasitic in terrestrial or aerial vertebrates. In the eggs of these parasites this function is effected by the protective coverings of the eggs, rather than by the osmoregulatory system. In the parasitic nematodes the ability to withstand desiccation (cryptobiosis) is well developed but the physiological mechanisms involved are not understood.

There is only a small literature concerning ion and water balance in parasites, probably due to the small size of most parasites and their inability to survive satisfactorily in culture. In the parasitic Protozoa, contractile vacuoles occur in some, but not all groups. It is not certain whether they carry out an osmoregulatory role since there are practical difficulties in determining with any accuracy very small fluxes of water and ions across the cell membrane. Some degree of osmoregulation has been observed in a few species, e.g. *Balantidium, Crithidia fasciculata, Entodinium, Ichthyophthirius multifiliis, Trypanosoma gambiense, Vahlkampfia calkensis.* Some of the intestinal amoebae develop contractile vacuoles when placed in hypotonic growth media and can remain viable over a wide range of osmotic pressures. The white-spot organism of fishes, *Ichthyophthirius multifiliis,* although unquestionably endoparasitic in habit, possesses many contractile vacuoles, but their function is far from clear.

The osmoregulatory role of the helminth protonephridial system has been examined in a small number of species, usually by adopting the procedure of measuring the flagellar beat rate of the flame cells when the worm is maintained in media of different ionic strengths and osmotic pressures. In general, these studies suggest that the protonephridial system is not involved in osmoregulation. Platyhelminths appear to be osmoconformers, so that the composition of the protonephridial canal fluid reflects the composition of the external environment.

In the miracidia of some digeneans, the beating of the flagella in the terminal organs is initiated, upon hatching, by light. In *Fasciola hepatica* no further change in the beating rate of the flame is discerned under different external conditions of osmotic pressure or ionic strength. In adult *Fasciola gigantica* that are placed into a hypotonic medium, the excretory bladder becomes distended and frequently empties its contents. This worm can tolerate a wide range of osmotic pressures without changing its weight through water fluxes. In other helminths, weight changes are a common response to the environmental conditions, e.g. *Cysticercus tenuicollis* loses weight in hypertonic media and gains weight in hypotonic conditions; *Hymenolepis diminuta* loses weight in hypertonic conditions and gains water, but loses salts in hypotonic salines. The protonephridial system of *H.diminuta* cannot control internal levels of Na^+, but does exert limited control over the concentration of K^+, although it is possible that this control resides with the surface membranes of the worm. Other tapeworms that have been examined (*Calliobothrium, Moniezia* and *Schistocephalus*) are also osmoconformers. It has been suggested that tapeworms with a free-living larval stage, such as the Pseudophyllidea and the Trypanorhyncha, would have a particular need to osmoregulate in the larval condition and this ability may or may not be maintained by the adult. *Lacistorhynchus tenuis* can

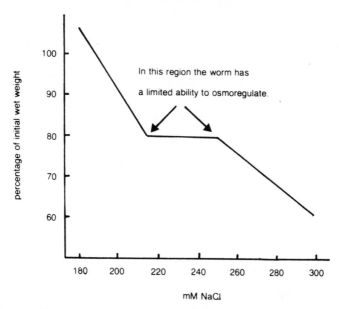

Figure 5.4 Weight changes in the cestode *Lacistorhynchus tenuis* related to the external concentration of NaCl. Redrawn from Read, C. P. and Simmons, J. E. (1963) *Physiological Reviews*, **43**, 263–305.

osmoregulate over quite a wide range of osmotic pressures (figure 5.4), as shown by weight changes in media of varying strength. Urea plays an interesting role in ormoregulation in the dogfish tapeworm *Callioboth-rium verticillatum*. This worm is freely permeable to urea and the addition of urea to a culture medium will prevent the inevitable uptake of water. This role of urea may not be surprising in view of its importance to elasmobranch fishes in maintaining a hypertonic blood. Urea may also play an important osmoregulatory role in other parasites of elasmobranchs.

The little information on osmoregulation in the Acanthocephala suggests that they are also osmoconformers—they do not possess any free-living transmission stages. The parasitic nematodes may possess limited powers of osmoregulation as shown by their ability to alter their body length in hypotonic or hypertonic media. Ionic regulation has been determined only for *Ascaris* which can control the pseudocoelomic fluid concentrations of calcium, magnesium, potassium and chloride ions to some extent. Juveniles of *Nippostrongylus* and *Ancylostoma* possess an

Table 5.1 Free amino acid concentrations in selected platyhelminth parasites.

Amino Acid	1 Hymenolepis diminuta	2 Schistosoma mansoni ♀ ♂		3 Dictyocotyle coeliaca	3 Entobdella soleae	3 Entobdella hippoglossi	3 Discocotyle sagitta	3 Diclidophora merlangi	3 Diclidophora denticulata
proline (% of total free pool)	4.3	6.6	5.4	63.9	37.3	72.5	8.0	72.3	41.3
glycine (% of total free pool)	10.9	8.3	8.4	2.0	31.8	9.7	17.5	2.0	4.8
total free pool (µmol/g dry weight)	56.9	45.3	40.8	1271.7	1441.1	1343.6	371.4	561.7	243.4

1. Chappell and Read (1973) *Parasitology*, **67**, 289–305.
2. Chappell (1974) *International Journal for Parasitology*, **4**, 361–369.
3. Arme (1977) *Zeitschrift für Parasitenkunde*, **51**, 261–263.

excretory ampulla which pulses at a rate that is inversely proportional to the osmotic pressure of their immediate environment.

Free amino acids and osmoregulation

Free amino acids in the tissue fluids are known to play a part in osmoregulation in some free-living invertebrates, proline and glycine, in particular, having a special role. Free amino acids may also be involved in the osmotic relationships of bacterial cells, and here again, proline features in a unique manner.

There is virtually no direct evidence that the free amino acid pool of parasitic organisms is concerned with ormoregulation. It is interesting to note, however, that the free pool of many ectoparasitic monogeneans is considerably higher than that of endoparasitic helminths. Monogeneans contain rather large amounts of free proline, and some also possess a large clycine pool (table 5.1). It is interesting to speculate that these free pool acids may be involved in osmoregulation; unfortunately there is no information on the size or contents of the free amino acid pool of freshwater monogeneans for comparison.

SUMMARY

1. The contractile vacuole is present in only some of the parasitic Protozoa. Its function may be osmoregulatory rather than excretory, but there is little direct evidence.

2. The protonephridial system of platyhelminth parasites may be of importance for the excretion of metabolic waste products. Ammonia, urea and amino acids have been identified as constituents of the protonephridial fluid in one species of tapeworm. Parasitic platyhelminths are mainly osmoconformers, so that the osmoregulatory role of the protonephridial system may be of limited significance.

3. Most parasites excrete ammonia, but some produce urea or uric acid. The external surfaces of many parasites are probably the major routes by which these toxic waste products are lost.

4. The osmoregulatory role of the free amino acid pool is suggested by the high levels of glycine and proline in some monogeneans.

CHAPTER SIX

REPRODUCTION

Introduction

This chapter contains a very brief account of the processes of asexual and sexual reproduction in parasites, from Protozoa to Nematoda. Limitation of space precludes a detailed description of the physiology of reproduction so that subjects such as sexual attraction, mating, spermatogenesis and oogenesis, development of the fertilized egg, egg shell formation and hatching of the mature egg, can only be treated in the most superficial of ways, or in some cases neglected altogether.

The reproductive biology of most parasites has not been studied in any great depth and, while the details of the morphology of reproductive systems are well known, relatively little physiological information is available.

Asexual reproduction

Asexual reproduction is achieved by the splitting or budding of a single cell or an entire organism, and therefore does not involve the fusion of two haploid gametes, with a subsequent meiosis. Asexual multiplication is a feature of many parasitic Protozoa, all of the Digenea and a small number of the Cestoda. In many of these parasites there is an alternation of sexual and asexual phases during the life-cycle, sometimes occurring in the same host.

In the Protozoa there are at least five distinguishable asexual processes that occur. *Binary fission* is by far the most common form of asexual multiplication and involves the parent cell dividing into two daughter cells. The division is preceded by a nuclear mitosis; the plane of division is transverse in the Ciliophora and Sporozoa and longitudinal in the Mastigophora. *Multiple fission* involves the secondary division of daughter cells before they become separated from the parent cell. It is a feature of the gregarines and trypanosomes and gives rise to spheres or

94

A.

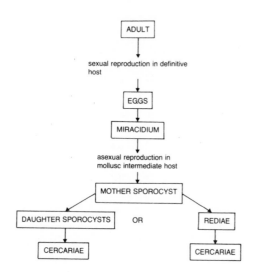

B. THE ASEXUAL STAGES OF CESTODA

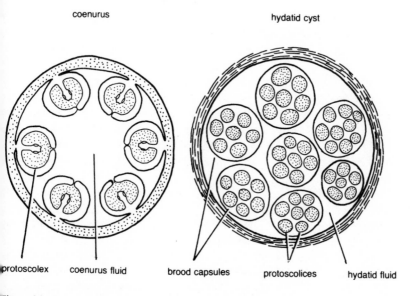

coenurus

hydatid cyst

protoscolex   coenurus fluid   brood capsules   protoscolices   hydatid fluid

Figure 6.1 Asexual multiplication in helminth parasites. The Monogenea, Acanthocephala and Nematoda to not have asexual stages in their life-cycles.
B. redrawn from Smyth, J. D. (1976) *Introduction to Animal Parasitology*, 2nd edition. Hodder and Stoughton.

chains of daughter cells. A third form of asexual reproduction is *schizogony*, a form of multiple budding, in which the parent cell (the schizont) produces daughter cells (merozoites) first by undergoing one or more nuclear divisions, and then by budding off many uninucleate daughters. Schizogony is typical of the Sporozoa, such as the Haemosporina, (e.g. malarial parasites). *Endodyogeny*, in which two daughter cells are budded internally within the parent cell, is only known to occur in some sporozoans, such as *Toxoplasma* and related forms. A fifth type of asexual reproduction, *single budding*, occurs in free-living but not parasitic protozoans.

Asexual multiplication in the helminths occurs exclusively by *internal budding* or *polyembryony*. In the Digenea, the asexual phases are found only within the snail intermediate host, and involve the sporocyst and redia stages. During the intramolluscan phase of development of a digenean there may either be two generations of sporocysts, i.e. mother and daughters, or the initial sporocyst (arising from the penetrated miracidium) may give rise to a redia, within which are developed daughter rediae—the precise pattern depends upon the species involved. Cercariae are developed asexually within daughter sporocysts or daughter rediae. The intramolluscan phase of the digenean life-cycle is highly proliferative, since one miracidium may produce as many as several million cercariae by asexual development involving sporocysts or rediae.

The sporocyst is generally a hollow, elongated organism, with an anterior birth pore and a well developed excretory system, but lacking a gut. At maturity, a mother sporocyst (derived directly from a miracidium) will be replete with growing daughter sporocysts or rediae; these emerge through the birth pore and migrate within the snail, usually to the hepatopancreas (digestive gland). Daughter sporocysts are morphologically similar to the parent sporocyst, and produce, by internal budding of germinal tissue, large numbers of cercariae, which escape through the birth pore when "mature". The redial stages are not unlike the sporocyst but possess a mouth and gut. Asexual multiplication in the Digenea involves the division of *germinal masses* within these larvae; these masses contain large numbers of germ cells, which proliferate during development. In the mother sporocyst germ balls are formed, each of which will become a daughter sporocyst—many hundreds of daughters can be produced by the division of this germinal tissue. Daughter sporocysts and rediae, likewise contain germinal tissue from which embryonic cercariae are produced. Cercarial production by individual sporocysts or rediae may continue, uninterrupted, for several months, or even perhaps years, as has been recorded in one case in a laboratory infection.

There is some argument concerning the exact nature of asexual multiplication in the Digenea. Some authors regard this form of multiplication as parthenogenesis (ameiotic or meiotic), while others consider it to be regenerative budding or polyembryony. It has even been suggested that the intramolluscan stages of digeneans reproduce sexually and that sporocysts and rediae are not larval forms at all. These views have not been substantiated and the debate is really philosophical and adumbrated by semantic problems. Nevertheless, however one regards the germinal lineage of the intramolluscan stages of digeneans, a clear function emerges, that of enormously increasing the numbers of infective individuals. This may be an adaptation to the great wastage that must occur once cercariae are liberated into the external environment (figure 6.1a).

In the cestodes, larval development involving asexual budding occurs in only a few groups (Diphyllobothriidae, Hymenolepididae and Taeniidae). The majority of tapeworms form only a single larva from each egg; of the many types of larval stage, only a small number are capable of proliferation by asexual budding. External budding occurs in two larval types, the urocystis (a budding cysticercoid) and the urocystidium (a budding strobilocercus). Internal budding is the more common pattern and is typical of the polycercus larva (*Parictotaenia paradoxa*), the coenurus larva (*Taenia multiceps*) and the hydatid larva (*Echinococcus granulosus*). Larval scoleces are budded from the cyst wall in the polycercus and coenurus, but they develop within brood capsules in the hydatid cyst, budding from the wall of the capsule (figure 6.1b). Asexual budding in the cestodes is a highly effective way of increasing the number of individuals during the stay in the intermediate host, e.g. hydatid cysts can grow continually, reaching a diameter of several hundred millimetres and containing several million larval scoleces, each a potential adult tapeworm. Coenurus larvae normally possess a less dramatic reproductive potential and contain only a few hundred larval scoleces.

Sexual reproduction

Protozoa

Sexual reproduction (figure 6.2) in one form or another occurs in the Sporozoa, the Ciliophora, the Opalinata and the Hypermastigida (insect gut parasites). One problem of sexuality in the Protozoa concerns the differentiation between fusion of gametes and fusion of individual organisms, such as occurs in conjugation; these processes are often difficult to distinguish. Where recognizable gametes are formed, they may be morphologically identical (iso-gametes) or dissimilar (anisogametes), e.g. microgametes ("male") and macrogametes ("female").

In the Sporozoa, an alternation of sexual and asexual phases frequently takes place, with gamogony leading to gamete production and sporogony leading to the asexual production of sporozoites (schizogony is an additional asexual phase that is found in some sporozoans). The

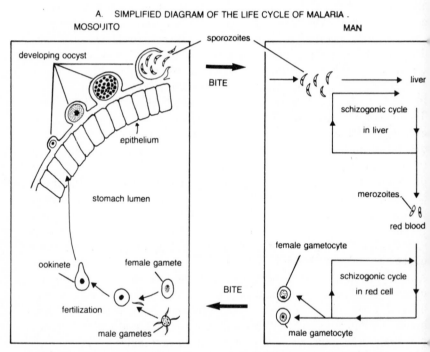

Figure 6.2 Patterns of sexual and asexual reproduction in the parasitic Protozoa.
A. Sexual reproduction takes place within the gut of the mosquito; the asexual phases occur in the liver and red blood cells of the bird or mammal final host and are termed schizogony. Modified from Schmidt, G. D. and Roberts, L. S. (1976) *Foundations of Parasitology*. Mosby.
B. Asexual schizogony precedes gamete formation, but both asexual and sexual phases take place in the gut of the same host. Modified from Smyth, J. D. (1976) *Introduction to Animal Parasitology*, 2nd edition. Hodder and Stoughton.
C. The complete life-cycle takes place within the gut of the insect. Based on the life-cycle of *Lipocystis* and redrawn from Grell, K. G. (1967) *Research in Protozoology*, **2**, 148–213.

B. SIMPLIFIED DIAGRAM OF THE LIFE CYCLE OF A COCCIDEAN

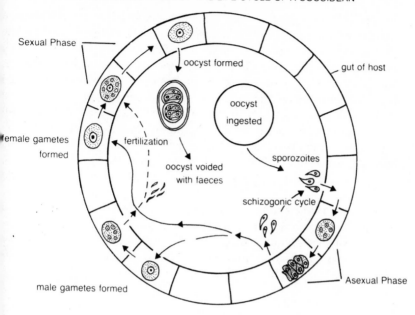

Sexual Phase

oocyst formed

gut of host

oocyst ingested

Female gametes formed

fertilization

oocyst voided with faeces

sporozoites

schizogonic cycle

male gametes formed

Asexual Phase

C. SIMPLIFIED DIAGRAM OF THE LIFE CYCLE OF A GREGARINE

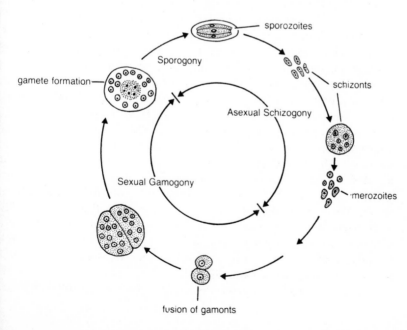

sporozoites

Sporogony

gamete formation

schizonts

Asexual Schizogony

Sexual Gamogony

merozoites

fusion of gamonts

major division of the Sporozoa into Coccidia and Gregarinida is based upon overt differences in sexual reproduction; in the gregarines gamete production occurs in both "sexes", while the coccidians form only microgametes, that resemble metazoan spermatozoa. In the Opalinata which are typically parasites of the posterior gut of amphibians, sexual reproduction involves the fusion of anisogametes which are releasesed from cysts produced only during the breeding season of the host. Fertilization takes place within the newly infected tadpole. In the flagellates, particularly those that inhabit the gut of insects such as termites, sexual reproduction involves the fusion of flagellated gametes. This process is under hormonal control and related to the moulting cycle of the host.

Where two different hosts are involved in the life-cycle (e.g. in the plasmodia) sexual reproduction will take place in the invertebrate host, while asexual sporogony and schizogony take place in the vertebrate. By contrast, asexual multiplication in the trypanosomatids occurs in both invertebrate and vertebrate hosts.

Monogenea

All monogeneans are hermaphrodite and there is usually a common genital opening for both male and female systems (figure 6.3A). Cross-fertilization, i.e. between two different individuals, is probably the normal method of reproduction, but self-fertilization may also occur.

The male system consists of testes, which vary in number and disposition, seminal ducts for sperm transport, various glands and a copulatory organ ("penis" or cirrus). The female system tends to be variable in structure from group to group, but basically contains a single ovary, extensive vitelline glands, various ducts, including a vagina which may have one or several openings to the outside, and, in the Polyopisthocotylea only, a genito-intestinal duct, whose function is unknown. The ootype is a muscular extension of the oviduct, surrounded by Mehlis' gland, a mucous and serous gland. The function of Mehlis' gland is not known. It was originally believed to contribute shell material to the developing egg, but this function is now thought to reside exclusively with the vitelline glands. Monogeneans are normally oviparous, laying shelled eggs into water, but one family, the Gyrodactylidae, is viviparous, bearing one, or frequently several developing embryos, one within the other.

Monogenean eggs are operculate, with filaments at one or both ends.

The egg shells are composed of tanned sclerotin, formed in the vitellaria and hardened by the quinone tanning system. The majority of eggs are liberated into water where they are presumed to sink to hatch, though the filaments may allow them to attach to a new host and hatch there; the latter is thought to be a rare occurrence. The period of incubation of the egg varies with species, e.g. it may be as short as three days, as in *Dactylogyrus* at 22–23°C, or it may take many weeks, as in *Dictyocotyle* at 10°C. The process of hatching of the monogenean egg has not received a great deal of attention and the roles of temperature, light, pH, pCO_2, pO_2 and salinity have not been determined. Light is an important factor for activating the developing larva within the egg and this may facilitate hatching, whereby the free-swimming oncomiracidium emerges through the opening of the operculum.

Digenea

The majority of digeneans are hermaphrodite (figure 6.3B) and probably reproduce sexually by cross-fertilization. Only in two families, the Schistosomatidae and the Didymozoidae, are separate males and females found.

The male system consists of paired testes, vasa efferentia, a single vas deferens, a seminal vesicle, an ejaculatory duct, and a cirrus enclosed within a sac. Spermatogenesis is typical of the phylum Platyhelminthes and sperm are stored within the seminal vesicle until copulation takes place. The female reproductive system consists of a single ovary and oviduct, a seminal (sperm) receptacle, paired and extensive vitellaria, an ootype surrounded by Mehlis' gland, Laurer's canal, which is homologous with the monogenean vagina, and an extensive uterus. There is a common genital opening for male and female systems, usually situated in the anterior half of the body. The male system normally develops first (protandry). At copulation the cirrus of one partner is inserted either into genital pore of the other partner or into the opening of Laurer's canal.

The development of the digenean egg has been extensively studied and the process of egg shell formation is well understood. Ova are produced in the ovary and are liberated periodically into the oviduct. Simultaneously, vitelline cells—from the vitellaria—and spermatozoa—from the seminal receptacle—are released. Fertilization of the egg takes place within the ootype and then the vitelline cells release their stored, shell precursors to form a soft shell around the fertilized egg. The role of the secretions of Mehlis' gland is unknown, but they may have a nutritional

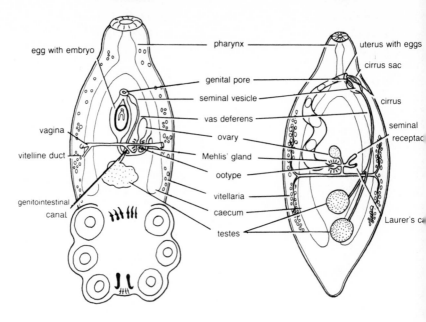

A. MONOGENEA   B. DIGENEA

egg with embryo   pharynx   uterus with eggs
                                                                    cirrus sac
                                    genital pore
                                    seminal vesicle   cirrus
                                    vas deferens
vagina   seminal
                                    ovary   receptac
vitelline duct   Mehlis' gland
                                    ootype
genitointestinal   vitellaria
canal   caecum   Laurer's ca
                                    testes

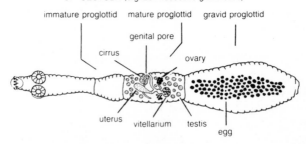

C. CESTODA (e.g. *Echinococcus granulosus*)

immature proglottid   mature proglottid   gravid proglottid

                        genital pore

        cirrus   ovary

        uterus   vitellarium   testis
                                    egg

Figure 6.3 Morphology of the helminth reproductive system.
A. redrawn from Schmidt, G. D. and Roberts, L. S. (1976) *Foundations of Parasitology.* Mosby.
B. redrawn from Erasmus, D. A. (1972) *The Biology of Trematodes.* Arnold.
C. redrawn from Smyth, J. D. (1976) *Introduction to Animal Parasitology*, 2nd edition. Hodder and Stoughton.
D. based on diagrams by Crompton, D. W. T. (1970) *An Ecological Approach to Acanthocephalan Physiology.* Cambridge University Press and Smyth, J. D. (1976) *Introduction to Animal Parasitology*, 2nd edition.
E. redrawn from Lee, D. L. and Atkinson, H. J. (1976) *Physiology of Nematodes*, 2nd edition. Macmillan.

D. ACANTHOCEPHALA   female   male

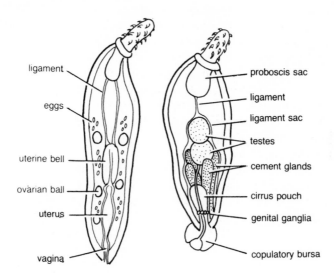

ligament

eggs

uterine bell

ovarian ball

uterus

vagina

proboscis sac

ligament

ligament sac

testes

cement glands

cirrus pouch

genital ganglia

copulatory bursa

E. NEMATODA   female   male

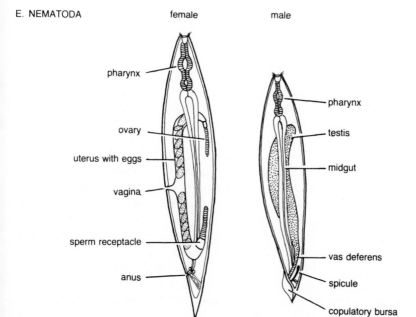

pharynx

ovary

uterus with eggs

vagina

sperm receptacle

anus

pharynx

testis

midgut

vas deferens

spicule

copulatory bursa

function. Shell formation proceeds as the fertilized egg is transported along the uterus. In a mature digenean the uterus is typically filled with eggs whose shells are opaque and brown due to the process of tanning. Quinone tanning is regarded as the most common method of providing the egg with a resistant covering, but this form of tanning may not be universal, e.g. *Fasciola* eggs contain proteins that are cross-linked by disulphide and dityrosine bridges rather than by quinones. Despite this, *Fasciola* does possess phenolases the enzymes which convert o-phenol to o-quinone during the process of quinone tanning. Thus the position remains somewhat enigmatic.

The tanned digenean egg is released into the external environment in host faeces or excreta. In most species the egg remains unembryonated until it is outside the host, but some do release eggs that contain well developed embryonic miracidia. The eggs are normally viable only if they are liberated into an aqueous environment since, despite the tanned shell, they lack the ability to resist dessication. In suitable conditions the egg develops to produce a single miracidium which is released from the egg during hatching—this depends upon the occurrence of appropriate stimuli such as light, temperature, gaseous conditions and salinity. Light appears to be a major hatching stimulus, stimulating the enclosed miracidium to become active or, perhaps, acting on light sensitive substances within the egg that facilitate the loosening of the operculum.

Cestoda

All cestodes, with the single exception of *Dioecocestus*, are hermaphrodites. Male and female systems are present in each proglottid (figure 6.3C), and the reproductive apparatus is therefore greatly replicated in an individual tapeworm. The Caryophyllaeidea, being unsegmented cestodes, contain only a single reproductive system. In the segmented worms the most mature proglottids are those that are furthest removed from the scolex and protandry marks the normal course of sexual development.

The male reproductive system is morphologically similar to its digenean counterpart, consisting of either discrete or diffuse testes, associated ducts, and a well developed cirrus within a pouch. The female reproductive system is also of the basic digenean pattern; the vagina is homologous with Laurer's canal and the morphology of the uterus is somewhat variable. Both self-fertilization and cross-fertilization occur, but little is known of the details of these processes for many species.

The cestode egg varies in its morphology according to life-cycle. In the

Pseudophyllidea, the egg is operculate with a thick surrounding capsule and matures outside the tapeworm, normally in water. The cyclophyllidean egg, by contrast, is surrounded by a thin non-operculate capsule, matures within the uterus of the gravid tapeworm proglottid, and hatches in the gut of the host (either an arthropod or a vertebrate, depending upon species). The fully developed embryo within the egg is termed the *hexacanth*, since it possesses six hooks. In the Pseudophyllidea the hexacanth develops into a coracidium larva upon hatching in water. The precise stimuli associated with hatching of the cestode egg will depend upon whether the egg hatches in the external environment or within the gut of the next host in the life-cycle. Pseudophyllidean eggs may require light to stimulate the opening of the operculum, though hatching of diphyllobothriid eggs has been recorded in the absence of light. Other factors that may be involved include temperature, pO_2 and salinity, but little precise information is available. There is rather more information on the factors that contribute to hatching of the cyclophyllidean egg, a two part process involving firstly, the breakdown of the embryonic membranes, and secondly, the activation of the hexacanth larva. The hatching mechanisms of taeniid eggs are regarded as somewhat different from those of other cyclophyllideans, due perhaps to the thin embryophore of the latter in contrast with the thicker embryophore of the taeniids. *In vitro* studies on hatching of taeniid eggs show that digestive enzymes, such as pancreatin and pepsin, and bile salts bring about the breakdown of the egg membranes, while an increased ambient temperature and bile salts appear to be responsible for larval activation. The eggs of other cyclophyllideans will hatch in saline solutions and do not require enzymes or bile salts—usually mechanical disruption of the egg membranes is an important factor for hatching.

Acanthocephala

Separate sexes and considerable sexual dimorphism are features of the Acanthocephala (figure 6.3D); the female worm tends to be rather larger than the male, and is thought to be longer lived, though the latter is a matter of some controversy.

The male reproductive system is completely enclosed within a ligament, which attaches anteriorly to the proboscis sac. The system contains two testes, a sperm duct, cement glands and a copulatory bursa. In the female worm, the reproductive system is in contact with the ligament, but is not enclosed by it. The female system consists of the ovary and ovarian balls, the uterine bell, uterus and vagina. At copulation spermatozoa gain access to the pseudocoelom, which contains the eggs, by way of the vagina, uterus and uterine bell. Copulation involves the insertion of the male bursa into the vagina of the female and after copulation the male plugs the vaginal opening with secretions that emanate from the cement glands. It seems improbable that female acanthocephalans can mate more than once. In the pseudocoelom, the spermatozoa fertilize the eggs, which are contained in the ovarian balls. The fertilized eggs are then liberated from the balls and continue their development within the pseudocoelom, entering the uterus when mature. The uterine bell acts as a sorting device. allowing passage of "mature" eggs, but keeping "immature" eggs in the pseudocoelom to complete their development. Ripe eggs are released from acanthocephalans at varying rates, e.g. *Polymorphus minutus* can liberate 2 000 eggs per day, while *Macracanthorhynchus hirudinaceus* releases up to 260 000 eggs per day; the prepatent period is from three to nine weeks depending upon species. The eggs, released into the external environment along with the faeces of the host, are surrounded by a protective covering consisting of three to four layers, which contain keratin, chitin and proteins. The shelled embryo, or acanthor, can withstand a range of environmental conditions and will only hatch after being ingested by the arthropod intermediate host. The egg, therefore, acts as a resting stage in the life-cycle which may remain quiescent for extended periods. Hatching in the arthropod gut is dependent upon pH and pCO_2, as well as on digestive enzymes, such as chitinases, to disrupt the shell membranes.

Nematoda

The great majority of nematode parasites have separate sexes (figure 6.3E), and exhibit varying degrees of sexual dimorphism, but hermaphroditic and parthenogenetic species also occur.

The male reproductive system consists of a single testis, a seminal vesicle, a vas deferens and a cloaca, but it is thought that the system was originally double. Spicules, which are hardened, extrusible, cuticular structures used to open the vulva of the female during copulation, are present in the males of most species, as either single or paired structures. They are located at the posterior extremity of the male. The female reproductive system contains paired ovaries, oviducts and uteri, which connect with a. single vagina, opening via the vulva. Frequently, part of the vulva is modified to form a muscular ovijector, responsible for extruding the eggs.

In the male system, spermatogonia develop either at the distal end of the testis, or, less commonly, along the entire length of the testis. In the female, the oogonia often develop in association with a cytoplasmic rachis, whose function is unknown. Subsequent growth and maturation of the oocytes takes place following their detachment from this rachis. Fertilization of the egg occurs within the seminal receptacle of the female. After fertilization the egg becomes surrounded by three shell layers; an inner ascaroside (lipid) layer, a thick chitinous layer and a thin outer layer. These layers derive from the egg itself, while the uterus may contribute a fourth layer, giving the egg a roughened and often sticky surface. In general, nematode eggs are remarkably resistant to chemical damage and dessication, due, in part, to the ascaroside layer which is composed of esterified glycosides. The eggs of ovoviviparous nematodes, such as *Trichinella* and the filarial worms, are, not surprisingly, less complex in their structure since the larvae hatch *in utero*.

Hatching of the eggs of the parasitic nematodes takes place either in the external environment, to release a free-living larva, or within the gut of a host animal, either intermediate or definitive host depending upon the life-cycle pattern. In *Trichostrongylus* and *Meloidogyne*, both of whose eggs hatch in the external environment, hatching is brought about by the egg shell becoming permeable to water and rupturing due to increased turgor pressure. The larvae of *Heterodera* actively cut their way out of the shell. Environmental factors such as light, temperature, pO_2 and humidity may all play a role in providing suitable conditions for hatching. Eggs that hatch within a host gut respond to the action of digestive enzymes and the prevailing physico-chemical conditions, especially temperature and pCO_2. The factors that render one host suitable and another less suitable for a particular species of nematode, by affecting the hatching of ingested eggs, have not been well investigated.

Synchronization of parasite reproduction with host cycles

There are a small number of cases known in which the reproductive activities of a parasite are coordinated or synchronized to the reproductive or other cycles of its host. Such synchrony often serves to bring about the co-occurrence of infective parasites and juvenile animals of the host species. We have already noted that, amongst the Protozoa, the Opalinata, which parasitize amphibians, release their gametes only during the breeding season of the frog or toad, and fertilization occurs within the recently infected tadpole larva. A related ciliate genus, *Nyctotherus*, which also parasitizes the posterior gut of amphibians, switches from asexual to sexual reproduction during the breeding season of its host. Experiments with various hormone treatments, using both *Opalina* and *Nyctotherus*, suggest that the levels of host sex hormones directly affect the reproductive activities of these parasites. In the hypermastigid flagellates inhabiting the arthropod gut, sexual reproduction is initiated by the moulting hormones of the host, e.g. in termites, the hypermastigid gut fauna is lost with each moult, so that sexual reproduction of the parasite is synchronized to the moult cycle; this is one method of reinfection where no cystic stage is produced. Reinfection is essential for the host since the flagellates produce the cellulases that digest the cellulose diet. Only one species of coccidian, *Coelotropha durchoni* in the polychaete worm *Nereis diversicolor*, shows reproductive synchrony with its host, though other sporozoans, including *Plasmodium*, *Leucocytozoon* and *Haemoproteus*, all synchronize to some extent with their hosts' breeding cycles.

Amongst the helminths, examples of reproductive synchrony are extremely rare and are limited to the monogenean *Polystoma integerrimum* in amphibians, and to synchrony of mitotic rate in the cestode *Diphyllobothrium dendriticum* with the circadian rhythms of the host. Many other cases will undoubtedly occur but these are yet to be described.

SUMMARY

1. Asexual reproduction involves binary fission, multiple fission, schizogony or endodyogeny in the parasitic Protozoa, and external or internal budding in the Digenea and Cestoda. In the helminths, asexual multiplication takes place exclusively within the intermediate host—a mollusc in the case of the Digenea and a vertebrate in the case of the Cestoda. This phase of the life-cycle is concerned with dramatically increasing the numbers of infective parasite larvae.

2. Sexual reproduction occurs in some protozoans and all metazoan parasites. The morphology of reproductive systems is briefly described, and the processes of fertilization, maturation of the egg and hatching outlined. Often there is an alteration of asexual and sexual phases in the life-cycle of a parasite, occurring in either the same or different hosts.

3. A small number of cases have been described where synchronization occurs between parasite reproductive activity and the reproductive or moulting cycles of the host.

CHAPTER SEVEN

PARASITE TRANSMISSION

Introduction

To the student of parasitology, the life cycles of parasites can often present a bewildering array of detail, with no obvious pattern appearing at first glance. Nevertheless there are certain events that are common in the life-cycles of all parasites, one of which is transmission to the next host. Transmission may occur more than once during a single life-cycle, should the parasite develop in one or more intermediate hosts. The adult or mature parasite completes its development and reproduces in the final, or definitive host, releasing eggs, larvae or other infective stages that must be transmitted to another host, whether it be of the same species (as in a direct life-cycle) or to hosts that are of different species (if the life-cycle is indirect).

The transmission of a parasite between consecutive hosts in the life-cycle can take place in one of three ways. First, transmission can occur when the potential host feeds on the egg, larval stage or on the intermediate host harbouring the larval parasite. Secondly, an intermediate host, or vector, may acquire the parasite and subsequently reinfect the definitive host, during feeding on host blood or tissues. Thirdly, transmission can often involve the active penetration of the host by free-living larval stages of the parasite. In an individual species of parasite, more than one of these modes of transmission may be employed at different stages of the life-cycle.

To help clarify the enigmas of parasite life-cycles, simplified schemes are shown below. In these diagrams, the stages outside the boxes are responsible for transmission. The variety of names used are summarized in table 7.1.

A parasite may inhabit widely differing environments at various stages in its life-cycle, and it will be adapted for life in each of these environments, however different they may be. Rather surprisingly, little attention has been paid by parasitologists to this ability of parasites to accommodate, both morphologically and physiologically to several quite

Direct life-cycle

e.g. many Protozoa (Coccidia, ciliates), almost all
Monogenea, some Cestoda, many Nematoda.

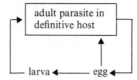

There is only one host in a direct life-cycle. The egg (oocyst in Protozoa) is released from the host, usually in the faeces, and is acquired by being eaten by the new host, or, if it hatches into a larva, either by ingestion or by penetration of the host by the larva.

Indirect life-cycles

Two hosts

e.g. many Protozoa (Sporozoa, flagellates), many Digenea, many Cestoda, most Acanthocephala, many Nematoda.

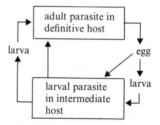

Passage through both hosts is mandatory for completion of the life-cycle. The egg may either be eaten by the intermediate host or it may hatch into a larva which actively penetrates or is eaten by the host (in the Protozoa the transmission does not involve eggs but various other stages of the life-cycle). The definitive host becomes infected either by eating the infected intermediate host or when the intermediate host feeds on host blood or tissues; penetration by the larva also occurs.

Three hosts

e.g. some Digenea (Strigeoidea), some Cestoda (Pseudophyllidea), some Acanthocephala.

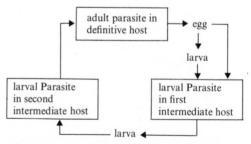

The addition of a third host to the life cycle merely amplifies the position outlined above. Usually both intermediate hosts become infected by active penetration of the parasite larvae; the final host normally ingests the infected second intermediate host for the life-cycle to be completed.

Figure 7.1 Simplified schemes of parasite life-cycles.

dissimilar environments, differing perhaps in ambient temperature, oxygen availability, salt and water content, pH and food content. The DNA content of a parasite will not alter during a single life-cycle, yet the parasite can perhaps adopt many forms and develop physiological attributes each unique to a particular phase in its life-cycle. Parasitic animals demonstrate this phenomenon to a much greater extent than do free-living animals. Little is known of the factors that bring about the "switching on" of some parts of DNA while other parts are "turned off". We presume that environmental stimuli are responsible for initiating such switching mechanisms via the action of operator genes which can activate the appropriate areas of DNA and inactivate the inappropriate areas. This will be discussed in more detail in chapter 8, when we consider the factors that control the establishment and development of parasites. Suffice it to say here that these putative environmental triggers, which operate throughout the life-cycle of a parasite, will play a particularly important role during the transmission process, because a parasite must be able to establish in a new host once that host has been located, i.e. its DNA must be switched to the appropriate regions once successful transmission has been accomplished.

Parasites typically produce large numbers of offspring in the form of either eggs or larvae, and this high degree of fecundity is usually regarded as an adaptation to the peculiar hazards of the life-cycle, involving time spent both inside and outside the host or hosts. Transmission between hosts probably represents a period of considerable loss in the life of most parasites, being a time when a great many individuals either fail to locate and enter a suitable host, or fail to establish successfully within the right host. The former will be discussed in this chapter, while the latter will be considered in chapters 8 and 10.

Mechanisms for locating the host

A variety of different stages are involved in the act of host location. These include oncomiracidia and adult worms in the Monogenea, miracidia and cercariae in the Digenea, coracidia in the Cestoda and various larval stages (usually the third larval stage) in the Nematoda. All of these individuals are active, free-living organisms that usually penetrate the host at its surface, or, in the case of the Monogenea, simply attach to the host surface. Research into the mechanisms employed both by parasite larvae and adults for host location involves behavioural studies, attempts to identify chemo-attractants and various biochemical approaches.

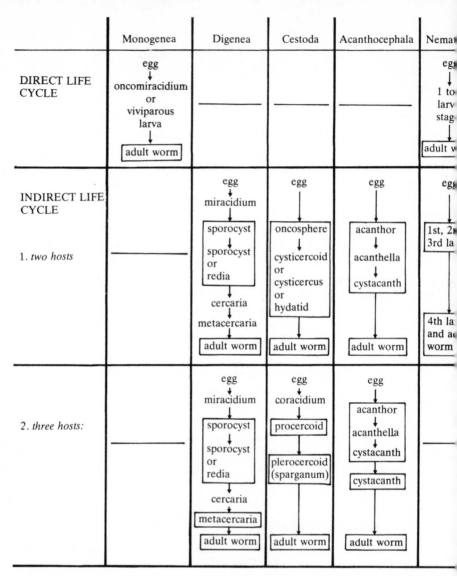

Table 7.1 Various stages in the life cycles of metazoan parasites. The stages represented outside the boxes are the free-living forms that are responsible for transmission to the next host. Each host is represented as a box within which the parasite may pass through a number of developmental larval stages, accompanied often by a dramatic increase in parasite numbers. Although there will be exceptions to the situations portrayed above, these patterns of life cycle and the nomenclature for each stage are valid for the great majority of parasites. Protozoan parasites, however, are excluded since they often have complex life cycles that do not readily fit the above scheme and the names of the various stages are not readily comparable.

Monogenea

The mechanisms by which monogeneans transfer from host to host have fascinated parasitologists for a very long time, but rather few species have been studied critically. Probably the best known example is *Entobdella soleae*, parasitic on the body surface of the flatfish *Solea solea* (the common European sole), through the work of G. C. Kearn almost a decade ago (figure 7.2).

Adult *Entobdella soleae* attach to the surface of the sole by a broad, cup-shaped *opisthohaptor*; the worm is quite mobile and capable of limited excursions, such as might be required for mating to be effected. Eggs are laid into the surrounding seawater and these hatch after 3–4 weeks, releasing the free-swimming larva, the oncomiracidium. This larva is small, about 0.25 mm long, and is ciliated in three distinct regions of its body surface. There are four pigmented eyes at the anterior end, each covered by a lens. Information on the host specificity of *E. soleae* strongly suggests that, under natural conditions, the parasite is quite definite in its preference for the sole and rarely occurs on other fishes. Is this brought about by active selection on the part of the parasite or do the oncomiracidia attach, but fail to develop in a wide range of fish species? These questions have been answered by the ingenious laboratory studies of Kearn, using isolated fish skin and scales.

When the oncomiracidia of *E. soleae* are offered a choice of isolated scales of its normal host in the presence of scales of related flatfish, such as *Microchirus* (*Solea*) *variegatum* (the thick-backed sole), *Buglossidium luteum* (the solenette), *Pleuronectes platessa* (the plaice) and *Limanda limanda* (the dab) (all of which co-occur around the British Isles) *Solea solea* scales are selected almost exclusively. It was noted that the parasite larvae settled, almost in every case, on the remnant of growing skin attached to the excised scale, rather than on the bony element of the scale itself, suggesting that fish skin might possess a specific attractant for the oncomiracidia. Attachment to isolated scales occurs in complete darkness, ruling out visual perception as a means of host location. The transfer of *E. soleae* from heavily infected sole, placed in a crowded aquarium tank with other species of flatfish, was limited primarily to recruitment of parasites by other sole and not by the other species of flatfish. Additional experiments showed that oncomiracidia will actively select and settle upon agar circles that have been impregnated with extracts of the skin of sole and this finding clearly implicates chemoreception as, perhaps, the major means by which *E. soleae* larvae

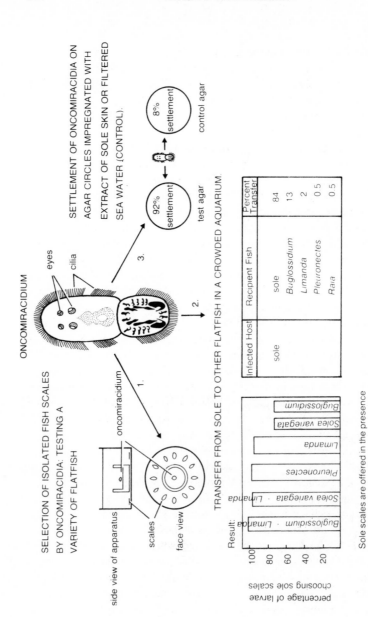

Figure 7.2 Host location and host specificity in the monogenean parasite, *Entobdella soleae*, of the flatfish *Solea*. The oncomiracidium is responsible for locating and settling on the most suitable fish, in this case exclusively the sole. Based on the studies of Kearn, G. C. (1967, 1970).

detect their host (see figure 7.2). It is evident that the mucoid secretions, from glands in the fish skin, are the principal attractants, since oncomiracidia are not attracted to sole skin lacking mucus glands, e.g. the cornea. It has been suggested that host-finding by chemotaxis in the Monogenea may be related to their evolution from a carnivorous, free-living ancestor, that found its prey by chemical identification.

Many other species of monogeneans are profoundly host-specific, but it is not known whether this phenomenon is brought about by the recognition of specific secretions of a suitable host or by other means.

Monogeneans will also transfer from host to host as adult worms when there is physical contact between host individuals and this is probably a common means of transmission in parasites of gregarious fishes. In fishes that do not shoal, parasite transmission may occur when the hosts make contact during their reproductive activities. These are areas where our knowledge is limited and they deserve further attention.

Digenea

The stages of the life-cycle concerned with transmission of digeneans are the miracidia, which hatch from the egg and penetrate the molluscan intermediate host (usually a gastropod), and the cercariae, which are released from the mollusc after a period of several weeks development and penetrate the next host, which may either be another intermediate host or the definitive host. The cercariae may, alternatively, encyst on vegetation as a metacercarial stage and reach the final host by the ingestion route, e.g. *Fasciola hepatica*. The majority of miracidia actively penetrate the mollusc, but some species must be eaten by the snail for continuation of the life-cycle, e.g. *Haplometra intestinalis* and *Glypthelmins quieta* are eaten by the snail *Physa gyrina*, then they penetrate the snail gut wall and invade the tissues.

The miracidium is a ciliated larva with sense organs that consist of eye spots and a variety of lateral papillae, which contain a nervous supply. Most miracidia respond emphatically to the physical elements of their environment, and they are thought to resemble their mollusc hosts in this regard. They are usually phototactic, thermotactic, thigmotactic and geotactic and will often respond to alterations in pO_2, pCO_2 and pH. Responses such as these probably function to bring the miracidia into close physical contact with molluscs. Once in the vicinity of a suitable snail, chemical attraction may facilitate the final stages of host location, although chemotaxis in miracidia is a controversial issue of long standing. While there is a substantial body of evidence implicating the release, by snails, of chemical attractants that are specific for individual species of Digenea and possibly mucoid in orgin, there exist in the literature a great many reports that fail to offer support for the hypothesis of chemical attraction and alternatively suggest that location of snails by miracidia is simply a process of trial and error.

Considerably varied experimental approaches have been adopted to test the chemical attraction hypothesis and apparatus employed includes four-armed mazes and Y-shaped choice vessels, collectively called chemotrometers. Other workers have studied miracidial locomotion under the flying spot microscope and yet others have preferred a strictly biochemical approach, using pure chemicals as potential attractants. Despite this wealth of experimental data, the controversy remains unresolved. Support for the chemical attraction hypothesis comes mainly from studies on the miracidia of *Schistosoma mansoni* and *Paragonimus ohirai*. In experimental chambers, the miracidia of *S. mansoni* appear to recognize and swim towards whole snails, crushed snails and snail extracts that have been impregnated in agar blocks. It is not completely certain, however, whether the parasite larvae recognize only those snails in which development can continue or whether unsuitable snails are also detected and approached. Schistosome miracidia detect specific chemicals, such as short-chain fatty acids, certain amino acids and sialic acid impregnated in agar blocks, and will alter their swimming behaviour according to the nature of the attractant. Unfortunately, this approach to the study of chemotaxis, as pioneered by A. J. McInnis, has not been thoroughly pursued using other species of digeneans, yet it would appear to be a fruitful line to adopt. Short-chain fatty acids may be recognized by the miracidia of *Fasciola hepatica*. There is, however, no unequivocal evidence that snails release specific substances that will attract only those digeneans which can grow in them. The chemical evidence available at present suggests that snail mucus contains a number of distinct components, any of which might be involved in attraction.

There are many reports in the literature that tend to refute the hypothesis of chemical attraction by snails, including studies on *Fasciola hepatica*, *Fascioloides magna*, *Schistosoma mansoni*, *Austrobilharzia* and *Zoogonoides* (see table 7.2). The age of miracidia can affect their ability to locate the host, but the mechanisms involved are not known, e.g. studies on *Megalodiscus temperatus* have demonstrated that 2–5 hours old larvae are more successful in locating the snail (*Helisoma trivolvis*) than either younger or older larvae.

Table 7.2 Selected references on miracidial location of snail hosts.

Species of digenean Miracidia	Host snail Species	Authors
Data supporting the chemical attraction hypothesis		
Schistosoma japonicum	*Oncomelania nosophora*	Lieper and Atkinson (1915)
Schistosoma mansoni	*Biomphalaria glabrata*	Kloetzel (1958, 1960), and Etges and Decker (1963)
	Helisoma anceps *Bulinus* spp.	
Paragonimus ohirai	*Assiminea parasitologica* *A. japonica* *A. latericea miyazakii*	Kawashima *et al.* (1961)
Megalodiscus temperatus	*Helisoma trivolvis*	Ulmer (1970)
Data refuting the chemical attraction hypothesis		
Zoogonoides laevis	*Columbella lunata*	Stunkard (1943)
Schistosoma mansoni	*Biomphalaria boissyi*	Abdel-Malek (1950)
Schistosoma mansoni	*Biomphalaria glabrata*	Stirewalt (1951), Chernin and Dunavon (1962)
Austrobilharzia variglandis	*Littorina pintada*	Chu and Cutress (1954)
Gigantobilharzia huronensis	*Physa gyrina*	Najim (1956)
Trichobilharzia elvae *T. physellae* *Schistosomattium douthitti* *Schistosoma mansoni*	A variety of normal and abnormal hosts	Sudds (1960)

For specific references and further details see Cheng (1967), Smyth (1966) and Ulmer (1970)

On the precise nature of a possible chemical attractant for miracidial host location, snail mucus has been put forward most frequently as the likely candidate. Yet, as we have seen, there is no unequivocal evidence to support this notion. It may be, therefore, that other snail exudates, e.g. reproductive or excretory material, faeces or haemolymph, could be involved, but there is little experimental evidence either to support or refute this. It seems probable, yet unproven, that miracidial responses to the physical environment bring them close to snails and that chemotaxis only operates, if operate it does, at close quarters. What is clear is that most miracidia are more strongly attracted by light than by any other stimulus.

Host locating mechanisms employed by cercariae, the other free-living digenean larva, are also surrounded by controversy. These larvae swim by means of a tail and do not appear to use their cilia for locomotion; some cercariae do not swim at all but crawl along the substrate to find the host. In the swimming forms the tail is lost once contact with the host is made and penetration commenced. Emergence of cercariae from the snail in which they have developed appears to be under the control of such factors as light, temperature, humidity, pO_2 and pH. In the laboratory, stress (induced by crowding snails) will also stimulate cercarial emergence. Once emerged from the snail, the cercaria exists as a free-living larva for only a short time, up to 36 hours as a rule. The behaviour patterns of swimming cercariae, which represent the most common type, are characterized by alternating periods of swimming and resting activity; during the rest period they tend to sink, rising in the water on resuming swimming. This is aided by their positive phototaxis, mediated through pigmented eyespots. Some cercariae do not possess eyespots and are negatively phototactic. These are commonly crawling (rather than swimming) cercariae that infect bottom-dwelling animals. Superimposed upon the primary response to light are additional responses that assist in the process of host location, e.g. the cercariae of *Posthodiplostomum cuticola*, which penetrate and encyst in a variety of European freshwater fishes, respond to shadows, such as might be produced by a passing fish, and swim spontaneously in a swarm. Spontaneous swimming is also initiated by increased water turbulence and by contact with a moving object. There is no evidence for chemotaxis in *P. cuticola*, however.

In general, cercariae are positively phototactic and negatively geotactic, with the noteable exception of crawling forms, but there is doubt as to whether chemotaxis plays any significant role in host location. Early studies on *S. mansoni* and *Clonorchis sinensis* indicated that chemotaxis was unimportant, but more recent work suggests that chemical attraction may play a not insignificant role in host location by cerariae. Chemotaxis may occur in a small number of species, including *Gorgodera amplicava*, *Schistosoma mansoni*, *Schistosomattium douthitti*, *Trichobilharzia*, *Austrobilharzia*, and *Gigantobilharzia huronensis*. Although it is not clear whether schistosome cercariae actually locate the host skin chemotactically, there is ample evidence that the specific site for cercarial penetration is selected by responses to chemical components in the skin itself, perhaps released by the sebaceous glands. Fatty acids may attract some schistosome cercariae to the host, since it has been demonstrated experimentally that *Schistosomattium douthitti* cercariae fail to penetrate excised mouse ears if the skin is extracted with ether to remove the fatty acids. Penetration resumes if ether-extracted ears are rubbed with fatty acids. The bird schistosome, *Austrobilharzia ter-rigalensis*, requires contact with sterols, particularly cholesterol, for successful penetration, but these substances do not appear to be attractants for host location by the cercariae.

Cestoda

The only active, free-living larval stage in the cestodes is the coracidium of the Pseudophyllidea. This is a ciliated larva, which hatches from the egg after its release into freshwater within the faeces of the bird or mammal definitive host. The coracidium is infective to copepod crustaceans. It differs in two ways from miracidia and cercariae; first, it appears to swim exclusively by random movements, and secondly, it is eaten by its intermediate host, a copepod. As it presents itself as food to its intermediate host, transmission is a passive process.

Transmission in all other cestodes also occurs exclusively by passive mechanisms involving the ingestion of eggs or larvae by the host.

Nematoda

Many plant-parasitic and some animal-parasitic nematodes, such as the hookworms of vertebrates, possess active, free-living larval stages that are concerned with locating a new host. Chemical attraction may play a particularly important role in host location by many of the plant parasites. Parasitic nematodes, in general, are well endowed with sense organs, including the antero-lateral amphids and the postero-lateral phasmids, which are thought to be chemoreceptors. Little direct evidence is available to support this hypothesis, however.

Host location in plant-parasitic nematodes is accomplished by random movements that operate when the parasite larva is some distance from the target plant, and by chemoreception when the plant is near. Mechanical stimulation may also be important for successful host location. Considerable exploratory behaviour characterizes plant nematodes when searching for a suitable entry site once the plant is reached. By contrast with the initial randomness that is a feature of the plant nematodes, the hookworms possess behavioural patterns that are clearly adaptations for locating a warm-blooded (homoiothermic) animal in a terrestrial environment. Larval mammalian hookworms, such as *Ancylostoma* and *Necator*, tend to be inactive for much of the time and are only stimulated into activity by the increased temperature due to a passing homoiotherm. *Nippostrongylus* larvae seem to search actively for their rat hosts by characteristic waving movements of the head and once a host is located the larvae spiral rapidly down the hairs following a thermal gradient.

Nematode larvae are more sophisticated in their behaviour patterns than the equivalent transmission stages of platyhelminths. They respond to a wide range of stimuli, including light, heat, gravity, various chemicals, mechanical events and electrical fields.

Mechanisms for penetrating the host

Information on the processes of active penetration of the host epidermis by larval parasites is rather scanty and tends to be concentrated on a small number of well studied species. The stages concerned with host penetration are miracidia, cercariae, nematode larvae and some parasitic insects, the latter remaining outside the scope of this book.

Digenea

The miracidium, as we have seen, locates and penetrates a snail and then migrates through the host tissues, finally settling in the digestive gland (hepatopancreas) where it develops into either the sporocyst or redia stage. There is some indication that miracidia search the snail epidermis for a suitable penetration site, but most appear to penetrate at the site of initial attachment. The larvae adhere to the snail surface by means of the apical papilla, which is a mobile and suctorial organ. Penetration is thought to be effected by secretions from the glands that open at the papilla (figure 7.3). *Fasciola hepatica* miracidia possess three types of papillary glands. There is a single large, multinucleate apical gland and several pairs of lateral, uninucleate glands, termed accessory glands; additional glands may open at the base of the papilla. When the miracidium is firmly attached to a suitable host the secretions of these glands are extruded by muscular contractions of the entire body of the larval parasite. Miracidia of *S. mansoni* and *S. mattheei* possess similar papillary glands. It is postulated that the secretions of miracidia contain a mucoid lubricant to aid both attachment and penetration and that lytic enzymes are responsible for breaking down the host epidermis. At (or perhaps just prior to) penetration, some miracidia, e.g. *Fasciola hepatica* and *Fascioloides magna*, shed their ciliated epidermal plates, whereas other species keep these plates intact during penetration, e.g. *S. mansoni*. The relevance of these events is obscure at present.

Penetration of mammalian skin by the cercariae of digeneans, most

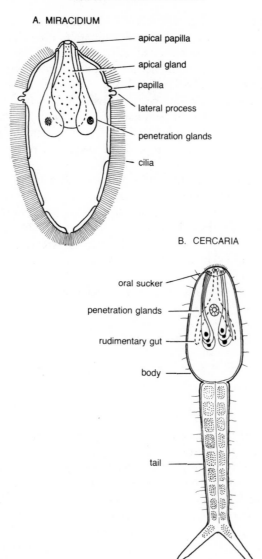

Figure 7.3 The gross morphology of skin-penetrating digenean larvae.

notably the schistosomes, has received considerable attention. Mammalian skin is a complicated structure made up of three distinct regions: an outer layer of keratinized dead cells (the *stratum corneum*), below which are living layers, the epidermis and the dermis, separated by a basement membrane. The layers of the skin vary in thickness and composition between different species of mammals, between individuals of the same species and even between different regions of the surface of the same individual. These factors may play an important role in determining both host specificity and the success of parasite establishment in any particular host.

There is some evidence that skin lipids are essential for stimulating penetration by schistosome cercariae, e.g. cholesterol is the major stimulant for the avian schistosome, *Austrobilharzia terrigalensis*, while mammalian schistosomes require free fatty acids. These lipids are thought to be important stimulants only after the cercariae have actually attached to the host skin, and they are probably not chemical attractants themselves.

The penetration of mammalian skin by the cercariae of *Schistosoma mansoni* has been categorized into six stages by Stirewalt (1966), and these may be summarized as follows:

1. Exploration of the skin.
2. Secretion of mucus for attachment.
3. Piercing and penetrating the *stratum corneum* through the muscular action of the cercarial body.
4. Abrasion of host tissue.
5. Mucus secretion to soften the *stratum corneum*.
6. Secretion of (putative) lytic enzymes.

Exploration of the host surface is much more a feature of cercarial than miracidial penetration, and it seems that cercariae, although capable of entering at any site, expend energy in locating the most suitable region, such as wrinkles, ridges, areas where dead cells are flaking off, or even hair follicles (gland-cell ducts are not used as sites of entry). Exploration is accompanied by the secretion of mucus from the cercarial post-acetabular glands, and this aids adhesion of the larva to the skin surface. The process of entry itself is part mechanical, brought about by the vigorous and abrasive actions of the cercaria, and part enzymic, due to the secretions of the acetabular glands, which are expelled by the body movements of the penetrating larva. On penetration, the cercarial tail is shed and the migratory larva is now called a *schistosomulum*, a term reserved for species of the genus *Schistosoma*. The cercariae of *S. mansoni* can penetrate a wide range of mammals but they do not survive to maturity in all the species entered. Thus, while the skin forms a distinct barrier, it is not the sole location in the host body at which host specificity may be determined.

The transformation from free-living cercaria to tissue-dwelling schistosomulum is a physiologically dramatic one, both in terms of temperature acclimatization (from 25°C to 37°C) and adjustment to a medium with a considerably increased ion content. A schistosomulum that has transformed from a cercariae for only a very short time will die if returned to fresh water at 25°C. Transformation is a facet of skin penetration, but it can also be accomplished experimentally, *in vitro*, by using a biphasic culture medium and also by allowing cercariae to penetrate excised mouse skin. The latter is a routine laboratory technique in the culture of schistosomes (figure 7.4).

Nematoda

Some members of the order Strongylida, such as the hookworms and related worms (Ancylostomatoidea, Strongyloidea, Trichostrongyloidea and Heligmosomatoidea) are transmitted to the vertebrate host, usually a mammal, by actively penetrating the skin as a free-living larval stage.

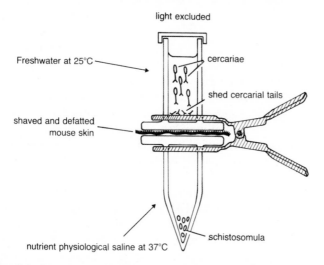

Figure 7.4 Apparatus used to transform schistosome cercariae into schistosomula *in vitro*. The all-glass flanged apparatus is incubated in a water bath for approximately 60 min, with excised mouse skin separating the two halves of the apparatus. Alternative methods of obtaining this transformation in the laboratory include mechanical shaking of cercariae to remove their tails or culture of cercariae in a biphasic medium.

The earliest evidence of skin penetration by hookworms was accidental, when in 1897, the famous parasitologist, Looss, became infected with *Ancylostoma duodenale* in error. Fortunately this mishap had no lasting effects on the pioneering studies of Looss in the field of parasitology. A variety of more controlled techniques has been subsequently employed to study the methods of skin penetration by larval nematodes. These included histological studies, histochemistry, tissue lysis studies, penetration through excised skin and behavioural approaches.

Early work on skin penetration by larval nematodes was carried out primarily *in vivo* and led, in general, to the conclusion that enzymes were secreted to effect entry across the skin barrier. More recent work has involved *in vitro* studies, using modifications of the Goodey "floating raft" technique, in which excised skin is stretched across a window in a sheet of cork that is floated in saline. Studies of this type have revealed that there is some variability in the mechanism of penetration adopted by larval nematodes. Initiation of penetration may come through a response to a thermal gradient, as shown by the hookworm, *Ancylostoma tubaeforme*, and entry occurs most frequently between dead cells of the *stratum corneum*. Some larvae exsheath (shed the cuticle of the previous larval stage) before penetration can commence, e.g. *Ancylostoma caninum* and *Necator americanus*, while others penetrate without exsheathment. The act of exsheathment, when it occurs, is thought to be initiated by skin lipids. Enzymes of parasite origin have been implicated in skin penetration in a number of species (*Necator americanus* and *Strongyloides fulleborni*), but it is by no means certain that all species rely on enzymes to effect entry across mammalian skin; equivocal evidence on this subject exists for *Ancylostoma caninum, A. ceylanicum* and *A. tubaeforme*. It is apparent that penetration may require a strict sequence of signals for the larva and this is exemplified by the fact that *A. tabaeforme* larvae cannot cross reversed skin (i.e. skin with the dermis presented outermost), *in vitro*.

Plant-parasitic nematodes possess a sharp anterior stylet which is used to pierce cell walls during feeding. The penetration of plant tissues, such as roots, appears to be effected exclusively by the mechanical action of the stylet, although there is some evidence that enzymes might assist with the process.

Circadian rhythms associated with transmission

Circadian rhythms, based on a 24-hour cycle of activity, are characteristic features of a number of parasites and are frequently associated with transmission either to the final host, if the life-cycle is direct, or to the final or an intermediate host, if the life-cycle is indirect.

Hawking (1975) has classified the circadian rhythms of parasites as follows:

1. Rhythms associated with synchronous cell division, e.g. malarial parasites.
2. Rhythms associated with the discharge of infective forms:
       a. from a definitive host, e.g. coccidia, pinworms, schistosomes.
       b. from an intermediate host, e.g. schistosomes.
3. Rhythms associated with parasite migration, e.g. trypanosomes, malarial parasites, microfilariae (larval filarial nematodes).

Circannual, or yearly, rhythmicity may represent yet another aspect of parasite transmission, particularly when associated with annual cycles of host occurrence.

Many species of malarial parasites are strictly periodic in their asexual reproductive cycle in avian or mammalian blood, with cell division, or schizogony, occurring every 24 hours (*Plasmodium knowlesi*), every 48 hours (*P. cynomolgi*) or every 72 hours (*P. malariae*). The circadian rhythmicity in malarial schizogony is closely related to the production and viability of the gametocytes, the stage involved in the process of transmission to the mosquito intermediate host, or *vector*. Male gametocytes of several species of *Plasmodium*, including *P. cynomolgi*, *P. knowlesi*, *P. berghei*, *P. chabaudi* and *P. cathemerium*, become ripe (shown by the occurrence of exflagellation) during a limited period of the day, which coincides with the feeding activities of the appropriate mosquito on the final host's blood, e.g. the male gametocytes of *P. knowlesi* are immature between 09.00 and 13.00 hours, mature between 21.00 and 05.00 hours and subsequently become moribund and die. It has been suggested that both schizogony and the time required for gametocyte maturation are, in each species, related to the biting times of the various mosquitoes involved in their transmission. It is not certain whether the female gametocyte possesses a similar circadian cycle of development. Experimentally-induced hypothermia in monkeys infected with malaria, has led to the conclusion that the synchrony of malarial schizogony is related, or *entrained*, to the daily temperature cycle of the host mammal, since prolonged cooling severely disrupts the parasite's circadian rhythm of development.

In malaria, then, circadian rhythms are associated with the requirement for the infective stage (the gametocyte) to be present in the peripheral blood at times when the transmission to the appropriate biting insect would be most likely to succeed. Other parasites possess similar rhythms which involve the active release of infective stages (eggs or larvae) from the host. The release of infective oocysts of coccidian parasites from the gut of birds is timed to coincide with the optimum period for transmission to another bird, e.g. in the case of *Isospora* in sparrows, the optimum time for transmission would occur in the late afternoon and early evening, since this is the time when roosting takes place, and many birds congregate; at other times of the day sparrows tend to be solitary in their existence. The life-cycle of *Isospora* is direct and infection is acquired by birds ingesting the liberated oocysts.

Mammalian pinworms (*Enterobius vermicularis* in humans and *Syphacia muris* in rats) deposit their eggs in the perianal region of the host by migrating from the rectum, their normal habitat, on a circadian basis. Migration of the female pinworms is associated with the daily lowering of the rectal temperature during sleep. The advantage of laying eggs during this period is probably to avoid contamination with host faeces, an occurrence that might reduce the chances of successful transmission, which takes place via scratching and the accidental ingestion of eggs. A similar cycle of egg release has been described for human infections of *Schistosoma haematobium*; a peak in the numbers of eggs present in the urine has been found to occur at midday. Hatching of schistosome eggs may also follow a circadian cycle.

Schistosome cercariae show strong preferences for liberation from the gastropod host at prescribed times of the day and this appears to be related to the optimum availability of the final host. *S. mattheei*, *S. mansoni*, *S. haematobium* and *S. bovis* cercariae tend to emerge during the early part of the day, while *S. japonicum* and *S. rhodaini* show a preference for emergence during the early evening and night. The emergence of *S. japonicum* cercariae at night suggests that this schistosome may have evolved as a parasite of nocturnal mammals and that man's involvement is perhaps accidental.

Many parasites commonly migrate within the tissues of their hosts and in certain cases these migrations are associated with transmission to an intermediate host; their function is to locate the parasite in the most suitable position within the host body for successful transmission to occur. Circadian periodicity in the distribution of microfilariae (larval filarial nematodes) in the mammalian blood system is a well known example of just such a migration. Hawking (1975) has examined the various microfilariae and recognizes four categories of microfilariae, according to their disposition within the host:

1. Numerous in the peripheral blood by night, rare or absent by day, e.g. *Wuchereria bancrofti* and *Brugia malayi*.

2. Numerous in the peripheral blood by day, rare or absent by night, e.g. *Loa loa.*
3. More numerous in the peripheral blood in the evening, e.g. *Dirofilaria immitis.*
4. Present in the peripheral blood over the entire 24 hour period, but more numerous in the afternoon, e.g. *W. bancrofti* in the Pacific region.

Microfilariae are transmitted by various species of mosquito and the feeding habits of these insects coincide with the time of appearance of microfilariae in the peripheral blood, e.g. day-biting mosquitoes transmit *Loa loa* and night-biting mosquitoes transmit *W. bancrofti.* When not present in the peripheral blood, microfilariae accumulate in the lungs, specifically in the pulmonary circulation at the junction between arterioles and capillaries. The process of larval nematode accumulation at this location appears to be related to the difference in oxygen tension (ΔpO_2) between pulmonary venous and arterial blood, e.g. in *W. bancrofti*, aΔpO_2 of 55 mm Hg is sufficient to induce accumulation of microfilariae in the lungs, and a drop of ΔpO_2 to 44 mm Hg will initiate migration of the nematodes to the peripheral blood. Artificial elevation of the pulmonary ΔpO_2, by taking exercise or breathing oxygen, will also bring about pulmonary accumulation of microfilariae. It is supposed, but yet to be substantiated, that the sense organs of the microfilariae are chemoreceptors, capable of detecting changes in pO_2 in the host blood. It is not known whether microfilariae possess an endogenous rhythm of their own, for, in the above example, they are clearly responding to an endogenous rhythm of host origin.

A similar circadian migration has been described for some species of trypanosomes inhabiting amphibians. One species, *Trypanosoma rotatorium* appears in the peripheral blood of *Rana clamitans* by day and accumulates in the renal blood vessels by night—this trypanosome is transmitted by leeches that feed by day. A second species of trypanosome, parasitizing the same host, shows the reverse pattern of migration and is thought to be transmitted by night-biting insects. It is not clear whether these migrations are due to an endogenous rhythm of the parasite or of the host, but they are presumably related to the light-dark cycle. There is no evidence for this type of migration in the mammalian trypanosomes, but some malarial parasites appear to be distributed differentially within the host blood system according to a 24 hour cycle.

Annual periodicity occurs in many parasites and there is some evidence that this may be related to transmission. It might well be regarded as wasteful for a parasite to produce and release infective forms if the potential host animals are not available for infection. Annual cycles, related to transmission, are evident in some malarial parasites, filarial worms, such as *Dirofilaria immitis* in dogs, *Onchocerca gutterosa* and *O. cervicalis* in cattle and horses, respectively, and in other nematodes, such as *Haemonchus contortus* in sheep. It is probable that circannual rhythms are, in fact, common in many parasites, particularly those inhabiting temperate regions of the world, where climate is seasonally demarcated and host occurrence varies considerably with season.

SUMMARY

1. Mechanisms by which the infective stages of parasites locate their hosts are discussed. Chemoattraction may be an important mechanism for larval monogeneans, but there is controversy concerning the host-locating mechanisms of digenean miracidia and cercariae. Some evidence supports the idea of chemoattraction, while other evidence suggests that alternative methods may be employed. It is possible that many parasites locate their hosts by trial and error or by accident. Behavioural responses to external stimuli by larval parasites are likely to bring the parasite and the potential host into close proximity.

2. The miracidia and cercariae of digeneans and many larval nematodes enter the host by active penetration through the surface. Miracidia enter at the site of attachment, aided by glandular secretions; cercariae tend to search skin for a suitable site for entry and they also penetrate with the assistance of secretions. Some larval nematodes penetrate the host skin barrier by secreting lytic enzymes, but it is not certain that all penetrating larvae enter this way. Plant-parasitic nematodes use a piercing stylet to effect entry.

3. Circadian rhythms of parasites are frequently associated with transmission to the next host. Such rhythms are described for malarial parasites, coccidians, schistosomes, pinworms, microfilariae and trypanosomes.

4. The host specificity of a parasite will be affected by its ability to locate and enter the appropriate host.

CHAPTER EIGHT

ESTABLISHMENT AND GROWTH
OF PARASITES

Introduction

A parasite must not only be successful in reaching the next host in its particular life-cycle, but must also establish itself, grow and mature if reproductive success is to be achieved. Sexual maturity will normally occur only in the final host. In the intermediate host (or hosts, should the life-cycle be indirect), the parasite may grow in size and may increase its numbers dramatically by asexual multiplication. Alternatively it may develop into a quiescent stage, such as the metacercaria of the Digenea, awaiting transmission to the next host, either by ingestion or death of the intermediate host.

In this chapter we shall consider the array of factors that permit parasite establishment in a potential host animal. A simplified sequence of events, depicting the barriers to invasion and establishment that confront a parasite during its infection of a single host, is illustrated in figure 8.1. This cycle will clearly be repeated twice, or even three times, in parasites that infect two or more hosts during a more complex life-cycle. The first barrier to invasion is primarily an ecological one and is outside the scope of this book. The second barrier represents the physical limitations to invasion, which have been discussed already in chapter 7. This physical barrier applies both to parasites that penetrate the host surface and to those that enter either the digestive system or hypodermically, via a blood feeding vector, and pertains to the conditions immediately encountered by the parasite once contact with the host has been made. The third barrier is a complex of physico-chemical and biotic factors that the parasite must be able to overcome or tolerate over a more extended time scale once access to the host body has been gained. These include the immune defence mechanisms of the host, which will be discussed in more detail in chapter 10. Here we shall consider the physico-chemical and interactive biotic factors that may allow or prevent parasite establishment, growth and maturation. Our attention will be focussed primarily on parasite development within the final host, since this has, in general, attracted the most attention from parasitologists.

Hatching mechanisms

Parasites that invade the host as cystic or egg stages enter the host passively during feeding. Such parasites excyst, or hatch, into larvae or pre-adults in the alimentary canal of the host, where they may either remain and develop to maturity, or from which they may migrate to other tissues in the host's body. The variety of stages that enter via the mouth are listed below.

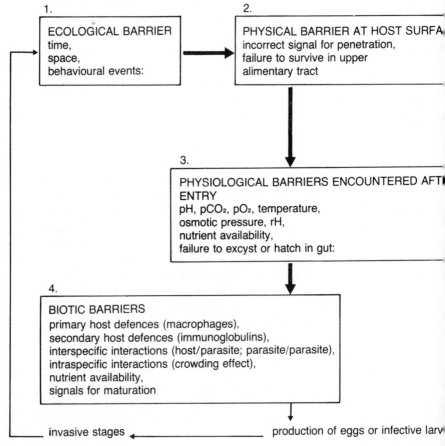

Figure 8.1 Principal factors determining the success of invasion, establishment and growth of a parasite.

Protozoans usually produce infective cystic stages that occur free in the environment, e.g. the oocysts of the Coccidia, and these are acquired when the host accidentally ingests them. Many digeneans form infective metacercariae, which are usually encysted and located within the tissues of an intermediate host; a few metacercariae inhabit the external environment (e.g. *Fasciola hepatica*). The cestodes form a variety of infective stages: these include eggs, that are liberated into the external environment; cysticercoids, cysticerci and hydatid cysts, that are encysted in the tissues of intermediate hosts; and the unencysted parasitic larvae, the procercoids and plerocercoids. Acanthocephalans are infective to the final host as the cystacanth larva, which is encysted in an invertebrate intermediate host. Nematodes are infective either through an embryonated second-stage larva within the egg or an ensheathed third-stage larva, the former occurring free in the external environment, the latter encapsulated within an intermediate host or free on herbage (trichostrongyles).

Almost without exception, the infective stage of those parasites that are acquired by being eaten by the final host have a reduced level of metabolism, i.e. they are quiescent. Possible exceptions to this general rule include the progenetic larvae, such as pseudophyllidean plerocercoids (Cestoda) and strigeid metacercariae (Digenea), which grow actively and, as larvae, are very close to the adult stage in their development, lacking only ripe gonads.

On ingestion by the host, the infective parasite must be able to withstand the rigours of life in the upper alimentary tract, particularly the stomach, but at the same time, it must be capable of utilizing a variety of stimuli to free itself from any protective covering and activate its hitherto quiescent metabolism. Precocious activation could prove lethal to the parasite, so a very delicate balance of factors must operate to ensure that hatching and activation take place in that region of the gut where the prevailing conditions can be tolerated by the liberated, immature parasite. This location is very often the duodenum in mammals and birds, but there are exceptions to this.

Protozoa

Excystation and activation of infective protozoan cysts has been studied in only a small number of genera, including *Eimeria*, *Isospora*, *Toxoplasma*, *Sarcocystis* and *Entamoeba*. Most of these investigations have involved *in vitro* systems and may, therefore, shed little light on the events that occur naturally. Excystation and activation can usually be accomplished, *in vitro*, by raising the ambient temperature to 37°C for mammalian parasites and to 43°C for avian forms, using a culture medium of neutral pH, containing reducing agents and with a low pO_2 and a high pCO_2. Bile salts may be an additional requirement for optimal excystation and activation. There is evidence that excystation and activation of protozoan cysts is a biphasic process, therefore, with *Eimeria* and *Entamoeba hystolytica*, initial activation of the encysted parasite is brought about by the elevation of temperature, the presence of reducing agents and a high pCO_2. Subsequent excystation is due to the activity of proteolytic enzymes, produced either by the parasite itself or by the host gut. Excystation in *Eimeria bovis* requires pretreatment with CO_2 and reducing agents, followed by trypsin and bile treatment. Mechanical disruption of the cyst walls may also be of importance in the initial processes, allowing the permeation of host enzymes. Rather unusually, the infective stages of *Histomonas meleagridis* are transmitted unencysted, within the eggs of parasitic nematodes (*Heterakis*), and in this way they are able to survive the potentially inimical conditions of the bird stomach.

In the Coccidia, the infective stages are sporozoites that are contained within sporocysts, themselves within the oocyst. Rupture of the oocyst micropyle is influenced by treatment with carbon dioxide in mammalian parasites, while in avian forms mechanical damage is thought to be more important. Excystation occurs when sporocysts are released from the oocyst and are exposed to bile and trypsin or chymotrypsin in the duodenum, thereby freeing the sporozoites into the gut. In many species of *Eimeria* and some species of *Isospora*, the sporocyst contains a structure, termed the *Stieda body*, which breaks down under the influence of host bile and trypsin to release the sporozoites. In other species of *Isospora* (*I. canis*, *I. arctopitheci*, *I. endocallimici* and *I. bigemina*) the sporocyst lacks a Stieda body and excystation occurs following the general breakdown of the sporocyst wall on exposure to proteolytic enzymes. In general, the process of excystation of coccidian sporozoites begins in the stomach or rumen and is completed in the upper small intestine. Coccidian inhabitants of the large intestine may, however, delay excystment until they reach a more posterior region of the gut, e.g. *Eimeria tenella*.

Digenea

The infective stage for the final host in many digeneans is the metacercaria, a quiescent larva that may be encysted either within an intermediate host or attached to vegetation, or alternatively, unencysted within an intermediate host, as is the case with some strigeid metacercariae, e.g. *Diplostomum spathaceum* and *D. gasterostei* within the eyes of many species of fish, and *D. phoxini* within the brain of minnows. The majority of metacercariae are enclosed within a cyst in the tissues of an intermediate host. The structure of the cyst varies widely with species, and thus the requirements for optimum activation and excystation may be very different. In those species that encyst on vegetation (Fasciolidae, Paramphistomidae and Notocotylidae), the metacercarial cyst wall is a much-thickened structure, capable of withstanding dessication and abrasion. In such cases excystation may be a complex process, requiring elevated temperature, bile salts and high pCO_2 to activate the larva, possibly stimulating it to secrete endogenous lytic enzymes. At the same time host proteases, such as pepsin in the stomach and trypsin in the duodenum, digest the cyst wall from the outside. In the case of *Fasciola*, the larva emerges from the cyst through a digested mucopolysaccharide plug in the cyst wall. Triggers for excystation are quite specific, since many metacercariae can pass through the alimentary canal of an unsuitable host and fail to excyst, yet retain their infectiveness.

The chemical structure of the walls of metacercarial cysts has not been adequately determined. The number of distinct layers that occur varies from one to four, depending upon species. *Parvatrema timondavidi* metacercariae are surrounded by a single layer of unidentified viscous material, and the metacercariae of *Posthodiplostomum* and Xiphidiocercariae are surrounded by two-layered cysts, composed of a layer of glycoprotein external to a layer of protein. Three-layered cyst walls are found in *Holostephanus luhei* and *Cyathocotyle bushiensis* (mucopolysaccharide, protein and host tissue), *Parorchis acanthus* (protein, glycoprotein and mucopolysaccharide) and in *Notocotylus attenuatus* (acid mucopolysaccharide and two protein layers). Four-layered cyst walls are a feature of the metacercariae of *Fasciola hepatica* (tanned protein, two layers of mucoprotein plus mucopolysaccharide, and an innermost keratin layer) and *Psilotrema oligoon* (mucopolysaccharide, protein, mucopolysaccharide and an inntermost protein layer). There is a large gap in our knowledge concerning the structure of the metacercarial cyst and the relationship between its structure and the optimal conditions required for activation and excystation of the enclosed larva.

With this wide variation in cyst structure, one would expect that the process of excystation in the vertebrate gut is equally varied in its biochemistry. Limited studies carried out *in vitro* suggest that this is, indeed, the case. In some species, such as *Parvatrema*, elevation of the ambient temperature is sufficient to cause excystation, while other species require pepsin pretreatment, followed by trypsin-bicarbonate or pancreatin treatment. It is not clear what contribution is made by the larval digenean itself in the process of excystation, but it may aid its liberation from the cyst by active physical movements or by secreting lytic enzymes. Host bile salts seem to be prerequisites for excystation in some species (e.g. *Fasciola hepatica*) but their role in general remains to be elucidated.

Cestoda

The infective stages of cestodes include eggs, cysticercoids, cysticerci, hydatid cysts, procercoids and plerocercoids, of which only the latter two are not covered by a protective cyst.

The eggs of many tapeworms hatch in a free aqueous environment, but those of the Cyclophyllidea hatch either within the vertebrate gut (e.g. Taeniidae) or in the gut of an invertebrate host. The cyclophyllidean egg possesses a thick, multilayered envelope that surrounds the larva (the oncosphere), the outer layer of which is called the *embryophore* (figure 8.2). In non-taeniid cyclophyllideans, the oncosphere is released from the egg, partly by the mechanical action of the host mouthparts, and partly by enzyme action. *In vitro* studies on hatching in hymenolepidid eggs suggest that bicarbonate, free CO_2, a neutral pH and proteases, particularly trypsin, are important factors. The taeniid egg is characterized by the presence of a considerably thicker embryophore. Hatching in these cestodes involves the enzymic disruption of the embryophore (by pancreatin in *Taenia psiformis*, and by pepsin in *T. saginata*) and activation of the oncosphere, which is then released from the enclosing oncospheral membrane. Bile salts are obligatory prerequisites for activation in these two cestodes.

Excystation of cysticercoids, cysticerci and hydatid cysts takes place in the vertebrate intestine following their ingestion along with the tissues of the intermediate host. Elevated temperature, a variety of host enzymes and bile salts have all been implicated in the processes of excystation and evagination of the larval scolex. In some, e.g. *Echinococcus granulosus*, pepsin alone is sufficient to cause excystation, but more usually trypsin plus bile or pancreatin plus bile are the minimal effective requirements *in vitro*. It may also be assumed that the prevailing physicochemical conditions are important contributory factors for both excystation and

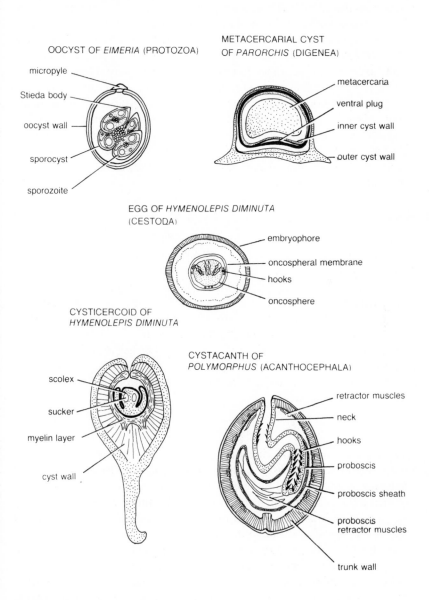

Figure 8.2 Some encapsulated, infective parasite larvae. Redrawn from Lackie, A. M. (1975) *Biological Reviews*, **50**, 285–323.

scolex evagination, but the roles of pO_2, pCO_2, pH, temperature and redox balance have not been examined thoroughly.

The plerocercoid larvae of the Pseudophyllidea are not encysted and are released from the tissues of the intermediate host during proteolysis and digestion in the upper intestine of the final host. These larvae are progenetic, resembling closely the adult worm in terms of size and somatic morphology. Sexual maturation in these larvae is probably induced by the elevated temperature in the bird or mammal gut in which they develop, along with the physico-chemical conditions that prevail. Because of their progenetic nature, the plerocercoid larvae of *Diphyllobothrium*, *Ligula* and *Schistocephalus* have proved relatively easy to cultivate to maturity *in vitro*.

Acanthocephala

Excystation and activation of acanthocephalan cystacanths has been examined in only three species: *Moniliformis dubius*, *Polymorphus minutus* and *Echinorhynchus truttae*. Activation of the larva occurs after mechanical or enzymic disruption of the thin envelope that forms the cyst. Evagination of the proboscis occurs only after the cystacanth has left the stomach of the host and it is thought to be stimulated specifically by bile salts and a high pCO_2. The exact role of bile is not fully understood—the cystacanths of *P. minutus* can be activated *in vitro* in the absence of bile salts, but the addition of bile salts to an activation medium increases both the percentage activation and the rate of activation of the cystacanths.

Nematoda

The eggs of some nematodes will hatch only after ingestion by a suitable host animal. In the case of the ascaridoid nematodes the process of hatching is initiated by the conditions in the mammalian or bird gut, inducing the secretion of various enzymes by the enclosed larva. These enzymes include chitinases, esterases and proteases, and it is their action that allows egress of the larva through a local, damaged region of the egg shell. Elevated temperature, a neutral pH, low pO_2 and high pCO_2 are responsible for triggering the activation of the larva within the egg.

Many animal-parasitic nematodes are transmitted to the final host, not as the egg stage but as third-stage larvae, ensheathed in the moulted (but not shed) cuticle of the second-stage. Prior to any development inside the host, exsheathment must take place in the host gut.

Ancylostoma, Haemonchus and *Trichostrongylus* all exsheath before development occurs and require specific stimuli to initiate this event. The major stimuli for exsheathment are high pCO_2, elevated temperature, the appropriate pH and the presence of reducing agents. Digestive enzymes of host origin do not appear to be involved in exsheathment in *Haemonchus contortus* and *Trichostrongylus axei*, whereas other species may require host proteases and bile salts for effective exsheathment to occur. Initiation of the exsheathment process is accompanied by the production of exsheathment fluid by the larval nematode. There is some controversy over the exact nature of exsheathing fluid, but it probably contains proteolytic enzymes, e.g. leucine aminopeptidase has been identified in the exsheathing fluid of *T. colubriformis*. The release of this fluid appears to cause the digestion of a localized, anterior region of the sheath, allowing the larva to escape into the gut.

Biochemical considerations of early establishment

The Role of bile salts

The role of bile salts as biochemical determinants of host specificity in gut parasites has been the subject of considerable speculation by parasitologists. Unfortunately there is, at present, insufficient evidence to lend support to this attractive and plausible hypothesis. Bile salts can be demonstrated to play important roles in the establishment of many of the parasites that enter the host through the alimentary canal. At least five distinct functions for bile are known:

1. Effects on membrane permeability.
2. Initiation of encapsulated larval activity.
3. Lytic effects on parasite surfaces.
4. Synergystic action with host digestive enzymes.
5. Metabolic effects upon both establishing and established parasites.

Because of the amphipathic nature of bile salts, i.e. each molecule possesses a hydrophilic and a hydrophobic end, they act as powerful surface- active agents, or *surfactants*. As such, bile salts can exert a profound effect on biological membranes, affecting both the lipid and protein moeities. The action of bile salts upon the membranes of parasite eggs and cysts increases their permeability, allowing the entry of water and digestive enzymes, through what was formerly an impermeable covering. In this way bile salts are thought to assist in hatching, excystation and exsheathment of many encapsulated parasites in the vertebrate gut. The story is more complex than this, since synthetic surfactants, such as Tween 80, do not necessarily act in the same way as naturally-occurring bile salts, and some may even inhibit the release of the parasite.

Bile salts stimulating movement of encysted larvae has been recorded in a number of species of parasite, e.g. *Eimeria* (Protozoa), *Cyathocotyle bushiensis, Echinoparyphium serratum, Fasciola hepatica, Holostephanus luhei* (Digenea), *Cysticercus pisiformis* (Cestoda), *Moniliformis dubius* and *Polymorphus minutus* (Acanthocephala), but neither the extent of this effect, nor the way in which it is effected is understood.

One of the mechanisms by which bile salts might influence the host specificity of either a permanent or a temporary parasitic resident of the vertebrate gut is suggested by the lytic effect of bile salts on certain parasites. This phenomenon has been demonstrated *in vitro*, with the protoscoleces of the hydatid organism, *Echinococcus granulosus*. Bile that is rich in deoxycholic acid, such as that of hare, rabbit and sheep, readily lyses the surfaces of the protoscoleces and thereby suggests a possible biochemical explanation for the unsuitability of these mammals as hosts for the tapeworm. Bile from suitable hosts, such as dogs, is typically low in deoxycholic acid. Although the extent of the lytic effects of bile have not been investigated, the limited data at present available do support the working hypothesis that bile may be one of the determinants of host specificity. Many parasite cysts rupture, releasing their larvae, under the direct influence of bile, in a region of the small intestine that is proximal to the opening of the bile duct. In the presence of unsuitable bile salts, therefore, parasite larvae may either fail to hatch or hatch but fail to survive due to the lytic effects of the bile. This hypothesis could readily be tested for a wide range of parasites by employing a relatively straightforward *in vitro* assay.

Bile may act synergistically with host digestive enzymes and this may assist in enzymic breakdown of parasite cyst walls. Mammalian trypsin and lipase activities are both increased in the presence of bile salts *in vitro*. Bile can additionally exert metabolic effects upon established as well as establishing parasites, e.g. experimental cannulation of the rat bile duct will both prevent the establishment of the tapeworm *Hymenolepis diminuta* and reduce the size and fecundity of already established infections. The effects of such experimental manipulations must be interpreted with caution, however, and additional evidence is required before any generalisations can be made on the metabolic effects of bile on parasites.

Metabolic changes during establishment

In many parasites, there is a typically complex life-cycle, involving free-living larval stages as well as parasitic larvae and adults. It is not unexpected. therefore, to find that metabolic differences exist in various parts of the life-cycle, though this has been investigated in only a small number of cases. Consequently, there is a marked lack of information on the stimuli, or triggers, that are responsible for bringing about the dramatic changes that must take place when, for example, a free-living stage of a parasite invades a mammalian host. From a biochemical standpoint there is one obvious and important difference between free-living, infective forms of parasites and fully parasitic adults; energy metabolism of the infective larvae tends to be aerobic, while the adult relies heavily, if not exclusively, upon anaerobic carbohydrate metabolism. Parasitic larval stages, such as sporocysts and rediae in the Digenea, resemble the adult worm and tend towards facultative anaerobiosis (table 8.1). Therefore, we note that the infective larvae of several species of nematode, the miracidia and cercariae of digeneans and, perhaps, the free-swimming coracidia of cestodes are all typically aerobic organisms. They possess a functional glycolytic pathway, Krebs cycle and β-oxidation of stored lipid, and their electron transfer system terminates in cytochrome oxidase. Parasitic larvae, such as sporocysts, rediae, procercoids, plerocercoids and acanthocephalan cystacanths, more closely resemble their adult forms in their metabolic pathways of energy metabolism, and are fundamentally anaerobic, lacking a complete Krebs cycle, relying upon glycolysis, and possessing branched electron transfer chains.

It is not at all apparent just what physiological and biochemical events control the dramatic switch from aerobiosis to anaerobiosis. During the process of invasion the parasite is seen to lose certain biochemical pathways and modify others. Isoenzymes, i.e. different proteins that catalyze the same chemical conversion, must be important during invasion, since there are often marked alterations in pH and temperature confronting a parasite on its journey from a free-living environment to the internal tissues of a host animal. It is not known whether the enzymes required for the adult stage are present, but inoperative, in the juvenile parasite, or whether they are synthesized *de novo* during and immediately after invasion.

Temperature and a high pCO_2 appear to be the most likely candidates for orchestrating the metabolic switch common to many parasites during invasion. Other factors, such as pO_2, pH, rH and osmotic pressure may also be involved. There is evidence that some larval digeneans (e.g. *Zoogonus rubellus* and *Himasthla quissitensis*) may be preadapted for the

Table 8.1 Metabolic differences between adult and larval parasites.

Parasite species	Summary of larval metabolism	Summary of adult metabolism
Ascaris lumbricoides	glycolysis- high pyruvate kinase activity; functional TCA cycle; β-oxidation of lipid; functional glyoxylate cycle; terminal oxidase is cytochrome oxidase	glycolysis to PEP—low pyruvate kinase activity; carbon dioxide fixation forming oxalacetate; mitochondrial reduction of fumarate to succinate; no TCA cycle; β-oxidation enzymes present but inoperative; no glyoxylate cycle; terminal oxidases are cytochromes o and a_3
Haemonchus contortus	functional TCA cycle	fixation of carbon dioxide
Schistosoma mansoni	*miracidia and cercariae* functional glycolysis and TCA cycles; cytochrome oxidase present; Pasteur effect demonstrable; probable β-oxidation of stored lipid *sporocysts.* utilize oxygen; cytochrome oxidase present; lactate excreted	typically anaerobic
Fasciola hepatica	*miracidia and cercariae* functional glycolysis and TCA cycles; large lipid stores *sporocysts* functional glycolysis; no cytochrome oxidase (present in rediae)	functional glycolysis; fixation of carbon dioxide; partial reverse TCA cycle; no lipid catabolism; branched cytochrome chain

higher temperatures encountered in the body of the final host, since there is a direct relationship between Q_{10} of the larva and the temperature of the final host. A number of parasites undergo metabolic changes as the ambient temperature is raised above $30°C$, e.g. *Nippostrongylus brasiliensis, Necator americanus, Strongyloides ratti* and *Schistocephalus solidus*. Conversely, parasites whose final host is a poikilotherm ("cold-blooded") must respond to factors other than temperature to initiate metabolic switching, but this has not been investigated in any detail.

Parasites are probably genetically rather complex, since the adult genome carries information for metabolic pathways that operate at other stages of the life-cycle. It would be interesting to speculate that the more complex the life-cycle of the parasite, the greater the quantity of functional DNA and RNA required. This hypothesis has not been tested, but evidence, from a single study, suggests that the genome of some cestodes and nematodes is more complex than that of equivalent free-living animals by a factor of between two and three.

Migration and site selection

Patterns of migration

Three readily discernible patterns of migration may occur once a parasite has invaded the host's body (figure 8.3). The first migratory activity serves to remove the parasite from its site of entry to the chosen target organ, where it will reside as a larva or as the adult stage. Such migrations are frequently accompanied, particularly in the helminths, by growth and development of the parasite, and conclude in the maturation of the worm. These are called *ontogenetic migrations*, since relocation and parasite development are closely related. Although we know very little about the events that take place during ontogenetic migrations of parasites, it is evident that a strict sequence of triggers exists, presumably emanating from the host and recognized by the parasite, that stimulate the sequence of normal development. This may be exemplified by the aberrant migrations that can occur when certain parasites invade a host in which complete development cannot occur, yet they may survive for some considerable time. This is not uncommon in skin penetrating forms, such as larval hookworms and larval digeneans. In the hookworms, penetration of an unsuitable host (e.g. dog worms entering man) may lead to the production of what are termed cutaneous or visceral *larva migrans*. These are migratory forms that fail to develop, lacking the necessary signals from the host, and they wander through the peripheral or deeper body tissues, often for extensive periods, causing pathological responses in the host during their migration and often

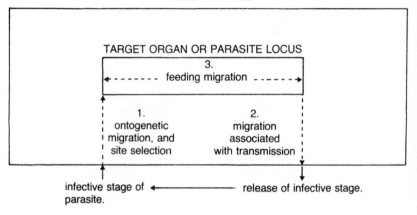

Figure 8.3   Migratory patterns of parasites.
1. Ontogenetic migrations. The parasite migrates from the site of entry to its preferred locus or target organ, usually undergoing development (growth and differentiation) during the journey. Ontogenetic migrations are both passive (i.e. via the host's blood) or active (i.e. through the host's tissues). Almost all metazoan parasites migrate ontogenetically.
2. Transmission migrations: Adult, sexually mature, parasites migrate from the target organ in which they reside to release infective stages (eggs or larvae), e.g. pinworms; alternatively the infective stages of the parasite migrate from the target organ occupied by the adult to the peripheral regions of the host body, e.g. microfilariae, schistosome eggs, some trypanosomes.
3. Feeding migrations: Adult or larval parasites migrate actively within their chosen target organ, associated with host feeding patterns; only known for *Hymenolepis diminuta*.

becoming encapsulated in organs and tissues. In the correct host these larvae would migrate via a strict route through the body, ultimately reaching the alimentary tract, where they become sexually mature and produce eggs. Another example is that of avian schistosome cercariae, which will penetrate the skin of man but never develop to the adult worm; a mild pathological condition, known as schistosome dermatitis, is produced at the site of entry. In man the schistosomules of avian schistosomes begin to migrate but this is brought to a halt in the dermal layers of the skin.

Ontogenetic migrations by parasites may result in either the adult stage reaching its preferred target organ or in the production of a quiescent larval stage that will await transmission to the next host (figure 8.4); digenean metacercariae are examples of the latter. It should also be noted that a great many parasites undergo ontogenetic migrations in their invertebrate intermediate hosts, e.g. trypanosomes and malarias in insects, digeneans in molluscs, cestodes in aquatic invertebrates and nematodes in insects. In these instances, successful transmission to the next host can only be accomplished after the completion of the ontogenetic migration, when the parasite is at the correct developmental stage and in a suitable location within its invertebrate host, e.g. tapeworm cysticercoids will be infective to vertebrates only after they penetrate the insect gut wall and complete their development in the haemocoel.

Parasite migration can be closely related to events that lead to transmission—this is typical of microfilarial nematodes and trypanosomes, and has been discussed at length in chapter 7.

A third type of migration has been characterized in the tapeworm, *Hymenolepis diminuta* and is thought to be related to the feeding patterns of the rat host. In worms that are eight days old, or older, there is a posterior migration in the intestine of the rat during the day and an anterior migration by night. This takes place in both single and multiple-worm infections when the rats are feeding by night and fasting by day, which is their normal routine. The pattern of migration can be reversed, experimentally, if rats are fed by day and denied access to food by night. This migration represents a real and considerable movement of tapeworms in the host's intestine, but the factors that initiate this migration have not been isolated. It has been suggested that either the passage of food or the gradually diminishing levels of free glucose in the intestine are responsible for stimulating the posterior phase of the migration, while the subsequent anterior migration might be along a gradient of glucose or some other substance, such as bile. Alternatively, the posterior movement could simply be passive, the worms being swept backwards, with the food, by peristalsis; once in the lower small intestine it may be that the worms recognize the unsuitability of their surroundings and actively migrate anteriorly to the duodenum. Certainly, they grow less well in the posterior intestine than in the more anterior regions. It is not known whether this type of diurnal feeding migration is a common occurrence in gut-dwelling parasites (it would be most likely to occur in parasites that were not firmly attached to the mucosa of the host intestine).

1. PARASITES THAT INHABIT THE HOST ALIMENTARY CANAL:

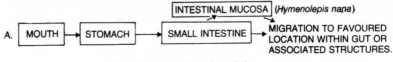

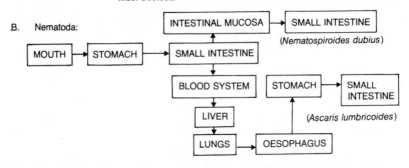

C. Nematodes that invade via a skin-penetrating larva: e.g. hookworms of man

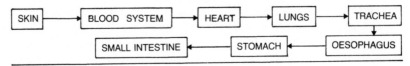

2. PARASITES THAT INHABIT HOST TISSUES:

A. Adult and larval Digenea that invade via a skin-penetrating cercaria:

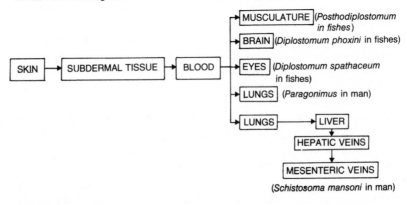

B. Cestodes that invade via the mouth:

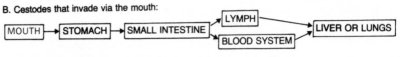

e.g. *Echinococcus granulosus* in man

Figure 8.4 Common patterns of ontogenetic migrations within vertebrate hosts.

Site selection and recognition

The great majority of invading parasites do not settle at the site of entry of the host but migrate to their preferred locus. Additionally, most parasites appear to be rather specific in their site requirements and rarely does one encounter ectopic parasitisms in the normal host. These facts strongly suggest that parasite migration from the site of entry to the final location is a directed event, involving the active participation of the parasite itself. The ubiquity of migrations associated with site selection should, ideally, allow us considerable insight into the mechanisms involved in the processes of site recognition and location. Unfortunately, this is not the case, and we know very little about the physiological signals which guide a migrating parasite to its preferred locus.

In the Monogenea, post-settlement migration is commonplace and many of the gill flukes of fishes change their position on the gills once the oncomiracidium has settled. Evidence of this has been obtained for *Tetraonchus monenteron, Diplectanum aequans, Diplozoon paradoxum* and *Discocotyle sagittata*. Analysis of the distribution patterns of *Discocotyle sagittata* and *Diclidophora merlangi* on the gills of trout and whiting, respectively, show clearly a deviation from the expected if settlement and establishment were random processes. These parasites actively select and migrate to specific areas of the gills, usually on the most anterior gill. A rather special case exists for the amphibian parasite *Polystoma integerrimum*. As a juvenile worm, the parasite inhabits the internal gills of the tadpole and, during the metamorphosis of the host, it migrates along the ventral, external surface of the host to the cloaca, from which it enters the urinary bladder, where it resides as an adult worm. Sexual · maturation of *Polystoma* coincides with gonadotrophin release by its host, ensuring that parasite eggs are released into the water at a time when larval frogs are abundant. Fertile, neotenic larvae of *Polystoma* are formed if the external gills of the tadpole are invaded, and a secondary cycle of development takes place (figure 8.5). The stimulus that controls the migration of juvenile *Polystoma* from the gills to the urinary bladder is probably endogenous, since larvae placed on metamorphosing tadpoles migrate first to the gills and then to the cloaca.

The majority of adult digeneans parasitize the alimentary tract, and its associated structures, of vertebrates. Little is known about their ability to select sites, but it may be assumed that this facet of their physiology is highly developed, as judged by their patterns of distribution within the host. Ingested digeneans will hatch or excyst in the upper small intestine where there is an abundance of proteases and bile salts, and from there they may migrate to other regions of the gut. Preferred sites in the Digenea include the stomach (*Otodistomum* in elasmobranchs, *Hemiurus* and *Derogenes* in teleosts), the duodenum (*Centrovitus*, *Glypthelmins* and *Loxogenes* in amphibians, *Mesocoelium* in reptiles, *Acanthoparyphium* in birds, *Nudacotyle* and *Alaria* in mammals), the posterior small intestine (*Lasiotocus* in teleosts, *Plagiorchis* and *Echinostoma* in birds, *Echinostoma* in mammals), the rectum (*Stephanostomum*, *Zoogonoides* and *Podocotyle* in teleosts, *Diplodiscus* in amphibians), the intestinal caeca (*Catatropis*, *Postharmostomum* and *Notocotylus* in birds). Other sites invaded, though

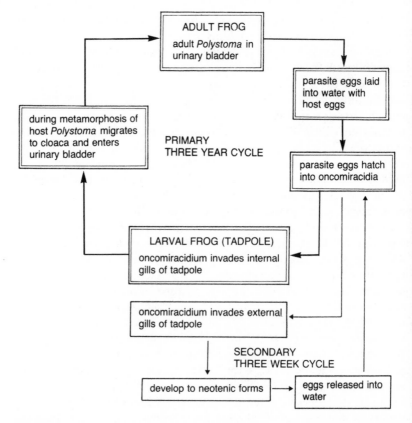

Figure 8.5 Site selection and migration in *Polystoma integerrimum* in Amphibia. The secondary, short cycle, leading to the production of neotenic larvae of *Polystoma*, arises only when the oncomiracidia invade the tadpole at its external gill stage; the oncomiracidia produced by this cycle then mature in the normal way, via the primary cycle.

less commonly, by adult digeneans include the mouth and buccal cavity (Plagiorchidae), the lungs (*Paragonimus*), the eustachian tubes (*Halipegus*), the pancreatic ducts (*Eurytrema*) and parts of the blood system (*Schistosoma*, Sanguinicolidae). From the above examples it will be obvious that migration to the final site can often be extensive, perhaps requiring the parasite to move against the flow of the gut contents or the blood. Although many of these migrations are well documented, little is known of the physiological mechanisms involved.

Some of the parasitic larval stages of digeneans are known to undergo extensive migrations within their intermediate hosts. This is typical for both the intramolluscan stages (miracidia to sporocysts and rediae) and the stages that inhabit vertebrates (cercariae to metacercariae). Larval strigeids, of the family Diplostomatidae, can migrate over large distances in their fish hosts. The fork-tailed cercariae may penetrate the fish at any region of its body, lose their tails and then migrate to specific organs that can be a considerable distance from the intitial site of penetration. The metacercariae of *Posthodiplostomum cuticola* encyst in the outer body musculature, *Diplostomum spathaceum* localizes exclusively in the lens of the eye, *D. gasterostei* in the retinal tissue and *D. phoxini* in the ventricles of the brain of their respective hosts. Studies on *Cercaria X*, which is probably *D. spathaceum*, demonstrate that migration can occur via the blood system, the lymphatic system and through the tissues; larvae that do not reach their target organ within approximately 24 hours of penetration are thought not to survive (figure 8.6).

Adult cestodes are all inhabitants of the vertebrate gut and their site selectivity within the alimentary tract is known for only a few species. The majority of tapeworms are located in the small intestine, or the equivalent region that lies immediately posterior to the opening of the stomach. Occasionally tapeworms occur in other regions, e.g. *Hymenolepis microstoma* in the bile duct of mice, and *H. fausti* in the caeca of ducks. Within the intestine, a tapeworm prefers a particular region for attachment, through this apparent preference can alter on a diurnal basis, as we have seen for *H. diminuta*. In the chicken gut, *Raillietina cesticillus* excysts close to the opening of the bile duct in the duodenum and, during growth and development of the worm over the next seventeen days, its position of attachment gradually shifts in a posterior direction, so that the fully mature worm inhabits a region 25 cm posterior to the opening of the bile duct. Transplanted *H. diminuta* will migrate to a preferred region of the small intestine regardless of the site of inoculation, but in this species, the notion of a preferred site is

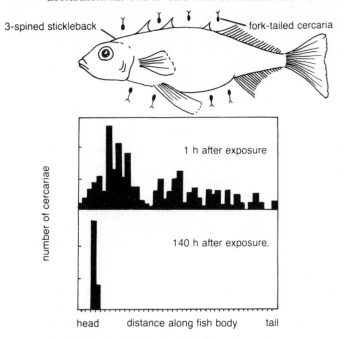

Figure 8.6 Host penetration and migration by the cercariae of a digenean whose metacercariae parasitize the eyes of freshwater fishes (*Cercaria* X = *Diplostomum* sp.).
The two histograms record the distribution of migrating parasite larvae within the body of the fish at two times following experimental exposure to cercariae. The numbers of migrating larvae were determined from serial transverse microscope sections of the fish. Larvae that failed to reach the eye within twenty-four hours are thought not to survive. Redrawn from Erasmus, D. A. (1959) *Parasitology*, **49**, 173–190.

complicated by the occurrence of a diurnal feeding migration. The fish tapeworm, *Proteocephalus filicollis*, inhabits the rectum of the three-spined stickleback while immature and the intestine when mature. All of the above three cestodes, then, exhibit ontogenetic migrations and a pronounced degree of site selection. In some other cestodes the site of attachment is determined by the morphology of the scolex in relation to the morphology of the host gut mucosa. This is illustrated by two tapeworms that cohabit the intestine of rays (*Raia radiata*): *Pseudanthobothrium*, which has a deep cup-shaped scolex, is restricted to the anterior three tiers of the intestinal spiral valve, where the gut villi are long; *Phyllobothrium* possesses a much flattened and foliaceous scolex and attaches only to the posterior tiers of the spiral valve, which have shorter villi.

Larval cestodes may be somewhat less specific in their choice of site, as they often inhabit the body cavity or musculature of the intermediate host. In the case of *Echinococcus granulosus*, the hydatid cysts develop in the first suitable site encountered after penetration through the gut wall, though this may be partly determined according to whether the blood or lymphatic system is used as a migratory route.

Adult acanthocephalans all inhabit the vertebrate gut, in which they are attached to the mucosa and submucosa by the armed proboscis. The proboscis itself is freely retractable so that the potential for movement and change of location within the gut exists. This has not been observed, however, except by implication in concurrent infections of *Moniliformis* with *Hymenolepis*, and *Neoechinorhynchus* with *Proteocephalus*. From observations on the location of acanthocephalans in natural infections it seems probable that each species has a preferred locus in the host's gut. Some species are found in the stomach, particularly in fishes (*Acanthocephalus lucii* in perch, *Acanthorhynchus borealis* in *Lota* and *Silurus*), but a large majority inhabit the upper intestine (*Pomphorhynchus*, *Neoechinorhynchus*, *Acanthocephalus ranae*, *Moniliformis dubius* and *Polymorphus minutus*), or the more posterior regions including the large intestine, the caeca in birds and the rectum in fishes. Larval acanthocephalans inhabit the body cavity of their invertebrate intermediate hosts, but it is not known to what extent they select specific sites within the haemocoel.

Nematodes parasitize a great many organ systems of invertebrates and vertebrates and each species appears to possess its own specific requirements in terms of sites selected. The much quoted study by Schad (1963) clearly demonstrates that several closely related species of pinworms, of the genus *Tachygonetria*, can coexist within the colon of turtles by virtue of their differences in site specificity, along with divergence in their nutrient requirements and feeding methods. Few other studies have examined site specificity in nematodes in such detail. It is evident, from the predilection of many nematodes for complicated patterns of migration around the host's body, that they must be able to recognize and respond to a variety of sequential stimuli concerned with their development, and that they are capable of actively choosing the site in which to reside.

The general subject of site selection in parasites has been largely overlooked by parasitologists, even to the extent that simple descriptions of the sites actually occupied by naturally-occurring parasites are often vague and imprecise. There is a whole field of parasite physiology lying fallow at present, which would be likely to reveal some very interesting aspects of host-parasite interaction. The most fundamental unanswered questions are:

1. What factors permit or prevent the establishment of a parasite at a particular locus in the host's body?
2. How do parasites recognize suitable environments for their continued development?

Invasion of host tissues

It is possible to recognize two separate processes of invasion in the life history of many parasites. The first invasion involves the entry of the parasite into the host's body, while the second invasion occurs when the parasite invades the particular tissues of the host in which it resides. Biphasic invasion, therefore, is typical of tissue parasites, but does not necessarily apply to those gut parasites that inhabit the lumen. Obvious examples of parasites with distinct invasion processes include the malarias and babesias that invade red blood cells, coccidians, nematodes and cestodes that invade the gut mucosa, schistosomes that invade the blood system and trypanosomes that invade nervous tissue.

Recent attention has focussed on the invasion of mammalian and avian red blood cells by malarial parasites, and of intestinal epithelia by coccidians. The accumulating evidence points to a high degree of specificity in this secondary invasion, and this feature may be important in determining the host specificity of certain parasites.

Malarial parasites (*Plasmodium*) invade the vertebrate blood stream as sporozoites, during the feeding of an infected mosquito on the host blood. The sporozoites do not remain in the perpheral blood of the mammal for long, but invade liver tissue where they multiply asexually, by schizogony, resulting in the production of merozoites. These forms are released into the blood stream and invade host red cells (erythrocytes), where a second schizogonic cycle takes place. The mechanisms by which the merozoites locate and then penetrate the erythrocytes are complex and far from completely understood. The merozoite is surrounded by a thick surface coat, which is lost during invasion of the red cell. The surface coat, carries a net negative charge, and is covered with fine filaments, that are involved in the initial attachment of the parasite to the host cell. It can be removed experimentally by treatment with proteases. At the anterior end of the merozoite is an apical protruberance which contains a complex of organelles (rhoptries, micronemes, polar rings and microbodies). These organelles characteristically disappear shortly after entry into the red cell is accomplished. The actual process of invasion is initiated when the merozoite attaches to the surface of the red cell by way of the surface filaments, and the apex of the merozoite is appended directly to the host cell surface (figure 8.7). This latter event causes local bending of the red cell membrane. Since the surfaces of both the parasite and the host cell are negatively charged, the initial attachment must be regarded as a "captive" process that overrides the natural electrostatic repulsion of the two cells. As entry proceeds, the red cell membrane invaginates, and this is thought to be due to the release of material from the rhoptries and micronemes. There is, therefore, no penetration, as such, of the red cell membrane and the engulfed parasite comes to lie within what is termed, a *parasitophorous vacuole*, bounded by the red cell membrane. The entire act of invasion may be very rapid, e.g. occupying as little as one minute in *Plasmodium knowlesi*. The mechanisms by which the parasitophorous vacoule is formed are not known, but a histidine-rich protein has been isolated from various stages of *P. lophurae*, and this induces cupping of red cell membranes at low concentrations, *in vitro*, and membrane lysis at higher concentrations. Whether such a protein is secreted by the apical organelles of the merozoite to facilitate invasion remains to be clarified. The sloughed-off parasite surface coat appears to plug the opening to the parasitophorous vacuole and may also aid the movement of the parasite into the red cell. Once invasion is completed, the merozoite transforms into the trophozoite which remains within the parasitophorous vacuole.

There is evidence that invagination of host cells is a common occurrence during tissue invasion by many protozoan parasites, and it may prove to be the typical method of invasion in both sporozoans and coccidians. *Babesia, Eimeria* and *Toxoplasma* all invade by the invaginative route. In these genera, the parasitophorous vacuole often breaks down shortly after invasion, resulting in an intimate contact between the parasite and the host cell cytosol. It has been alternatively suggested that *Toxoplasma* invades the host mucosal epithelium by inducing phagocytosis and avoiding subsequent digestion. However, inhibitors of phagocytosis, such as colchicine and cytochalasin B, do not interrupt invasion unless the parasite itself is treated directly, implying active invasion by the merozoite.

In the same way that the merozoites of *Plasmodium* recognize and attach to red blood cells, so *Eimeria*, among the coccidians, exhibits a marked preference for intestinal epithelial cells. So strong is the apparent degree of site specificity that sporozoites of *Eimeria* will develop only in

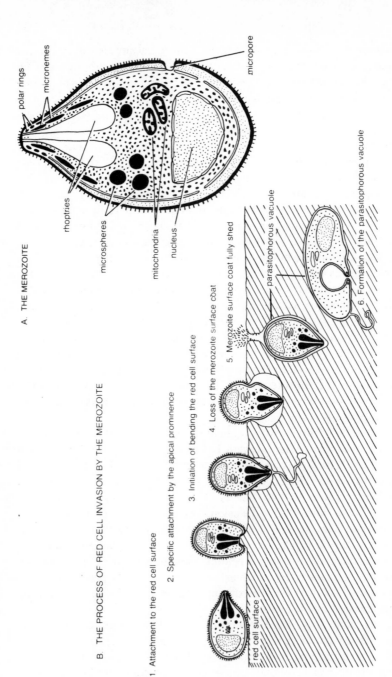

A. THE MEROZOITE

polar rings
micronemes
rhoptries
microspheres
mitochondria
nucleus
micropore

B. THE PROCESS OF RED CELL INVASION BY THE MEROZOITE

1. Attachment to the red cell surface

2. Specific attachment by the apical prominence

3. Initiation of bending the red cell surface

4. Loss of the merozoite surface coat

5. Merozoite surface coat fully shed

6. Formation of the parasitophorous vacuole

red cell surface

parasitophorous vacuole

Figure 8.7 The invasion of red blood cells by malaria parasites. Redrawn from Bannister, L. H. (1977) *Symposia of the British Society for Parasitology*, **15**, 27–56.

gut cells, regardless of whether the site of entry to the host's body is via intravenous, intramuscular, intraperitoneal or parenteral injection. Little is known about the invasion of gut epithelia *in vivo*, but there is a growing body of information based on *in vitro* studies. In culture, *Eimeria* sporozoites will invade and develop in cells from at least seventeen different hosts. Entry into these cells is rapid and appears to be an active process. The role of the apical organelles in this process is not clear. There is controversy concerning whether the sporozoites of *Eimeria* are phagocytosed by the host cell or whether they actively induce phagocytosis. Treatment of cell cultures with anti-phagocytic drugs suggests that the latter event occurs.

Recognition of the suitable host cell by a tissue parasite will depend upon surface receptors situated on the host cell membrane. There may even by receptors that are specific for attachment and others specific for invasion, since the merozoites of *Plasmodium* will attach to a variety of cell types, but will only invade erythrocytes. Invasion of red cells by malarial parasites is unaffected by neuraminidase treatment, but is reduced by treatment of red cells with proteases, concanavalin A and wheat germ agglutinin. Although the chemical nature of these surface receptors remains elusive, it is evident that they are labile entities that are readily lost or neutralized.

Tissue invasion by helminth parasites has been inadequately investigated and little is known about the processes of cell recognition or invasion. It is not known whether helminths invade host cells by the invaginative route or by direct penetration.

Factors inhibiting parasite growth and development

Crowding effect

Intraspecific crowding of parasites, within host tissues or within the gut lumen, may result in a reduction in their rate of growth, the maximum size attained and their fecundity. Only a small number of examples of the crowding effect have been described, yet it seems likely to be a common phenomenon amongst the helminths.

Probably the best described example of crowding is that of *Hymenolepis diminuta*. When this tapeworm is grown in carefully controlled infection densities in rats, there emerges a clear relationship between the intensity of infection and a reduction in growth rate, maximum worm weight and length, and egg production (figure 8.8). Any increase in the number of worms in an infection, however, has no effect upon the rate of maturity of *H. diminuta*, and egg production always commences on the sixteenth day after infection. There is, as yet, no unequivocal explanation for the crowding effect in this cestode. Competition for space, for oxygen and for essential nutrients, such as glucose, have all been proposed as likely candidates. Alternatively, the production of toxins or growth inhibitors by the worms may be responsible, and the immune responses of the host may be directly or indirectly involved. Crowding of *H. microstoma* in the mouse bile duct affects these worms in a similar manner.

The crowding effect has been observed in some species of parasitic nematodes, e.g. the length of adult *Skrjabingylus nasicola*, which inhabits the nasal sinuses of weasels (*Mustela nivalis*), is reduced in infections of increased intensity. *Ancylostoma caninum* infections of dogs are also subject to a crowding effect.

Interspecific interactions

Interactions may take place between cohabiting parasites of different species, and these can result in reduced parasite growth and fecundity, and also be accompanied by alterations in site selection. Concurrent laboratory infections of rats with *Moniliformis dubius* and *Hymenolepis diminuta*, result in a marked restriction in the distribution of each species within the rat's small intestine, as well as a reduction in size of both (the tapeworm shows the greatest size decrease in this relationship). A similar situation can be found in natural infections, e.g. co-occurring *Proteocephalus filicollis* and a species of *Neoechinorhynchus* mutually alter the site of each other in the gut of the three-spined stickleback, so that each species is restricted in its range as a result of the interspecific interaction. This is termed *competitive exclusion*, but the factors that bring it about are not known.

Other classes of interspecific interactions may exist between the parasite and its host, e.g. variations in the size and fecundity of parasites in relation to the size or sex of their host. Furthermore, a single species of parasite may grow and develop quite differently in a range of host species. Such interactions will involve a variety of physiological factors that include the immune response of the host, its immunological status, as well as the physico-chemical conditions that promote or limit establishment and growth of a parasite.

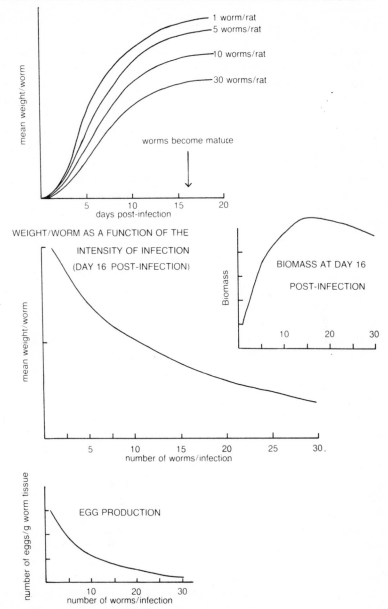

SPECIFIC GROWTH CURVES OF WORMS IN INFECTIONS OF INCREASING INTENSITY

1 worm/rat
5 worms/rat
10 worms/rat
30 worms/rat

mean weight/worm

worms become mature

5   10   15   20
days post-infection

WEIGHT/WORM AS A FUNCTION OF THE
INTENSITY OF INFECTION
(DAY 16 POST-INFECTION)

mean weight/worm

5   10   15   20   25   30
number of worms/infection

Biomass

BIOMASS AT DAY 16
POST-INFECTION

10   20   30

number of eggs/g worm tissue

EGG PRODUCTION

10   20   30
number of worms/infection

Figure 8.8 The "crowding effect" in the cestode *Hymenolepis diminuta*. Based on data from Roberts, L. S. (1961) *Experimental Parasitology*, **11**, 332–371.

Labile growth patterns

Members of the phylum Platyhelminthes possess a remarkable ability for degrowth and dedifferentiation under adverse nutritional conditions. This is characteristic of the free-living Turbellaria and the Cestoda, but it is not known whether it occurs in the Monogenea or Digenea. In the tapeworms, degrowth is termed *destrobilation*, whereby the majority of the proglottids are shed, leaving only the scolex and the neck region attached to the mucosal surface. Destrobilation can be induced in experimental animals by depriving the host of carbohydrate, and this has been demonstrated with *Raillietina cesticillus* and *Davainea proglottina*. Natural destrobilation has been observed in *Dibothriocephalus*, a parasite of bears, when the host's diet changes prior to hibernation. Natural destrobilation may be common in cestodes that inhabit animals which hibernate, where it would serve as an overwintering response by the parasite. Experimental deprivation of carbohydrate may induce a marked reduction in tapeworm size, without the obvious shedding of proglottids. This occurs in *Hymenolepis diminuta*, *H. citelli*, *Lacistorhynchus tenuis* and *Oochoristica symmetrica*, and it reflects the dependence of many cestodes on the availability of carbohydrate. Other tapeworms, e.g. *Schistocephalus solidus* and *H. nana*, do not possess such a high degree of sensitivity to depletion of dietary carbohydrate. Cestodes with a typically high rate of growth are more sensitive to a carbohydrate deficiency than those that grow more slowly.

A second feature of labile growth and development is found in the development of the hydatid cysts of *Echinococcus granulosus*. The protoscoleces within the cyst are capable of development and differentiation in one of two separate directions according to the environmental signals available. Should the cyst burst and release the protoscoleces into the body cavity or the tissues of the intermediate host, each protoscolex can become yet another cyst, in which new protoscoleces will develop; this is termed *secondary hydatidosis*. Protoscoleces that enter the alimentary canal of a suitable host will grow into mature tapeworms. A similar degree of lability in development is seen in the related cestodes *Taenia serialis* and *T. multiceps*. *In vitro* studies on *E. granulosus* suggest that the stimulus for strobilar development to the adult worm is contact with a proteinaceous substrate. Cultivation of protoscoleces in a liquid medium always results in the formation of secondary hydatid cysts, whereas provision of a punctured protein base induces strobilar development (figure 8.9). The ability of protoscoleces to become secondary hydatid cysts represents a particular problem for the surgeon, and he must exercise considerable care during extirpation of a cyst to prevent its rupture and the release of vast numbers of protoscoleces. Although it is not fully understood how environmental stimuli control the direction of development in *Echinococcus*, it is evident that a

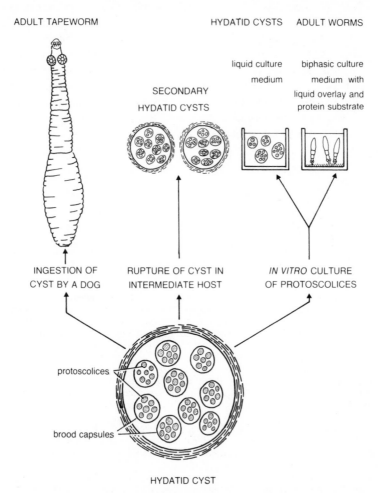

Figure 8.9   Development of the protoscoleces of *Echinococcus granulosus* (Cestoda) *in vitro* and *in vivo*. The protoscoleces will develop into either secondary cysts or adult worms depending upon the environmental conditions encountered. This is thought to be controlled at the level of operator genes. Based on data from Smyth, J. D. (1969) *Parasitology*, **59**, 73–91.

complex interaction between operator and repressor genes will take place in the switch from larva to adult worm, or larva to secondary larva.

Moulting in nematodes

A detailed consideration of growth and the phenomenon of moulting in the parasitic nematodes is outside the scope of this book. Suffice it to stress here that, because of their cuticular covering, all nematodes grow by distinct stages and moult, or shed the cuticle, between each growth stage. Commonly there are five growth stages, with four moults, between egg and adult, and in the parasitic nematodes the third-stage larva is normally infective to the host. In some species, the cuticle of the second stage is retained after moulting, forming a protective sheath around the larva; exsheathment takes place in the gut of the host.

SUMMARY

1. Hatching mechanisms of parasite eggs and excystation of encysted larvae in the alimentary canal of the host are discussed. Examples include protozoan oocysts, digenean metacercariae, cestode eggs and larvae, acanthocephalan cystacanths, and nematode eggs and ensheathed larvae. The roles of host enzymes, bile salts and the physico-chemical conditions of the vertebrate and invertebrate gut in the processes of hatching and excystation are considered.

2. The biochemical effects of host bile salts on establishing parasites are varied. It is thought that bile composition may play an important part in determining host specificity of many gut parasites.

3. When a parasite infects its host, there are often profound metabolic alterations to the parasite, including temperature acclimation and a switch from aerobiosis to anaerobiosis. The biochemical conditions that stimulate these changes are unknown.

4. After successful infection of the host, many parasites undergo migrations within the body of the host. These migrations may be ontogenetic, associated with transmission to the next host, or associated with host feeding patterns.

5. Migrating parasites show a high degree of site specificity within the host, and examples of site selection are described. The physiology of migratory orientation and site selection is poorly understood.

6. Many parasites invade the tissues of their host and settle within cells. The processes involved are described for malarial parasites and the coccidians.

7. Interactions between parasites of the same species or of different species may alter the growth and fecundity of such co-occurring parasites. Site selection may also be changed when parasites cohabit the same host.

8. Some platyhelminth parasites respond to adverse conditions within the host by degrowth. This phenomenon is not uncommon in the cestodes, where it is termed destrobilation. It may be related to poor nutritional conditions or to the action of the immune system of the host.

9. Labile growth and differentiation are a feature of some cestodes, such as the hydatid organism. The environmental conditions encountered by the larval tapeworms determine the pattern of development.

CHAPTER NINE

NERVOUS SYSTEMS, SENSE ORGANS AND BEHAVIOURAL COORDINATION

Introduction

This section contains a brief description of the nervous systems and sense organs that are typical of helminth parasites. No consideration of the parasitic protozoans will be given, though receptors are present on the surface of these parasites.

Morphology of parasite nervous systems

The platyhelminth nervous system consists of a cerebral complex of ganglia and commissures in the anterior region of the body, proximal to the pharynx. Entending anteriorly and posteriorly from the cerebral ganglia are bilaterally symmetrical nerves that innervate the tissues and organs, and which are joined by numerous lateral branches. The major deviation from the standard pattern that occurs in the parasitic platyhelminths is associated with the innervation of the organs of attachment, the posterior haptor in the Monogenea, the oral and ventral suckers in the Digenea and the scolex in the Cestoda.

The nervous system of monogeneans is basically comparable to that of the free-living Turbellaria. In some monogeneans an apparently primitive condition occurs whereby the cerebral ganglia are situated dorsal and anterior to the pharynx (e.g. Dactylogyridae and Tatraonchidae), but, more usually, the brain, composed of a pair of ganglia joined by a single dorsal commissure, lies immediately adjacent to the pharynx. In a few species, a pair of commissures join the cerebral ganglia, and completely encircle the pharynx. Of the three pairs of major nerve trunks that run posteriorly from the brain, (i.e. the ventral, dorsal and lateral nerves) the ventral pair are the best developed and innervate the posterior haptor, often developing ganglia in the haptor itself. Anteriorly, three or four pairs of nerves supply the head.

The basic pattern of the digenean nervous system differs little from that of the Monogenea; there is an additional nervous supply to the anterior and ventral suckers, which is often extensive. In the Cestoda, paired cerebral ganglia, joined by commissures, give rise to a pair of lateral nerve trunks, that extend the length of the strobila and are connected by commissures, which are particularly evident at the junctions between proglottids.

The nervous system of the parasitic platyhelminths is in the form of a nerve net, and contains unmyelinated nerve fibres, which are usually demonstrated, at the light microscope level, by the histochemical localization of acetylcholinesterases. There is little ultrastructural information on the platyhelminth nervous system, but there is no justification for regarding it as a degenerate system.

The nervous system of the Acanthocephala has been examined in detail in only a few species, e.g. *Polymorphus*, *Macracanthorhynchus*, *Bolbosoma* and *Hamanniella*. It consists of a cerebral ganglion, situated ventrally in the proboscis receptacle, and varying numbers of both paired and single nerves which supply the body wall, the musculature and the somatic organs. In male acanthocephalans, the lateral nerves enter the reproductive system, forming a pair of genital ganglia connected by a circular commissure.

Parasitic and free-living nematodes possess essentially identical nervous systems, consisting of a nerve ring of circum-oesophageal commissure, on which are located varying numbers of ganglia, one of which is dorsal, one ventral and two or more situated laterally. Anteriorly, six nerves extend to supply the mouth and head, while between six and eight main posterior nerves serve the body. The major nerve trunk, the ventral posterior nerve, is ganglionated at intervals and is paired along some of its length. At the posterior extremity the dorsal and ventral nerves divide and, on either side of the body, fuse with the lateral nerves to form a pair of lumbar ganglia. Sensory function is associated with the lateral and ventral nerves and motor function with the dorsal and ventral nerves.

Simplified, outline diagrams of nervous systems of helminth parasites are given in figure 9.1.

Sense organs

Monogenea

The major sense organs in the Monogenea consist of photoreceptive

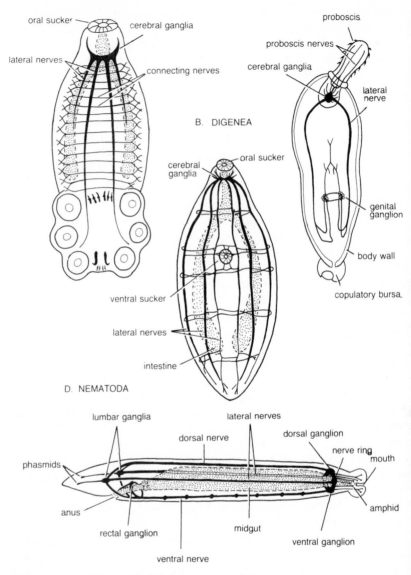

Figure 9.1 Morphology of the helminth nervous system.
A. redrawn from Bychowsky, B. E. (1957) *Monogenetic Trematodes.* American Institute of Biological Sciences.
B. redrawn from Erasmus, D. A. (1972) *The Biology of Trematodes.* Arnold.
C. redrawn from Hyman, L. H. (1951) *The Invertebrates: Acanthocephala, Aschelminthes and Entoprocta.* McGraw-Hill.
D. redrawn from Lee, D. L. and Atkinson, H. J. (1976) *Physiology of Nematodes,* 2nd edition. Macmillan.

eyes, ciliated, superficial or epidermal structures (sensillae) possibly concerned with chemoreception, tangoreception and rheoreception, and haptoral knobs, that may be touch receptors. While the morphology of these structures has been examined in some detail with both light and electron microscopes, relatively little is known of their true physiological role, due to the many difficulties incumbent in an experimental approach.

Although many monogeneans possess eyes, these organs tend to be more a feature of the oncomiracidium and are often absent in the adult. This may be indicative that photoreceptors are important in host location, but are not necessary to the adult worm once it is established on its host. Primitively (in the Monopisthocotylea), there are two pairs of dorsal eyes, each with a crystalline lens; the anterior pair are small and are directed posterolaterally, while the larger, posterior pair are directed anterolaterally. The arrangement in the Polyopisthocotylea is regarded as more advanced, there being only a single pair of eyes, either lacking a lens or with an oil droplet lens. In all monogeneans the body of the eye consists of a pigment cup, possibly containing melanin, and a nervous supply. The functional significance of eyes in the larval monogeneans is not clear, since there is little information of their behavioural responses to light. There is evidence of positive phototaxis in some oncomiracidia, but this may be a short-lived response (e.g. Dactylogyridae). Most larval monogeneans swim with their dorsal surface uppermost, so that dorso-ventral orientation may be one function of the eyes, at least during daylight swimming. There is evidence for the occurrence of additional, putative photoreceptors in monogeneans, based exclusively upon ultrastructural studies of larval *Entobdella soleae*. These photoreceptors are paired structures, located on the head of the oncomoracidium, close to the eyes. They contain many modified cilia, lamellate bodies and are richly provided with mitochondria. These receptors are morphologically similar to ciliary photoreceptors described in other invertebrates, but convincing evidence for their precise function has yet to be provided.

There is a variety of possible sensory structures associated with the monogenean surface, usually taking the form of papillae containing modified cilia. Their sensory function is deduced from histochemical techniques, particularly silver staining, but this deserves cautious interpretation. These sensory organs are located over much of the body surface and are concentrated in the head region and on the haptors. Electron microscope studies have revealed three types of sensory structure: single receptors (nerve bulb plus a cilium); compound, uniciliate receptors (several nerve endings, each with a single cilium);

and compound, multiciliate receptors (one or several nerve endings; each with many cilia). The single sensilla, the most common type of receptor, has been identified in a number of species (*Gyrodactylus, Entobdella soleae, Amphibdella flavolineata, Leptocotyle minor* and *Diclidophora merlangi*) and is thought to be a touch receptor. Compound receptors have been described in *E. soleae, Gyrodactylus, Acanthocotyle elegans* and *A. lobianchi*, and these are thought to be chemoreceptors. The haptoral knobs of larval *E. soleae*, with a rich supply of nerve endings but lacking in cilia, may be mechanoreceptors.

Digenea

Miracidia and cercariae are the free-living digenean larvae concerned with host location and penetration. We have already discussed these processes, in general terms, and examined the controversy concerning active host seeking as opposed to accidental discovery. Most of the studies on digenean sense organs have been made on these two larval forms.

Miracidia The morphology of miracidial sense organs is known for only a small number of species (*Schistosoma mansoni, Fasciola hepatica, Allocreadium lobatum, Heronimus chelydrae, Philophthalmus megalurus* and a species of *Spirorchis*). These sense organs are of several types: pigmented eye-spots, ciliated pits, ciliated nerve endings on the apical papilla, and ciliated nerve endings, both on the lateral papillae and between the ciliated body plates. The eye-spots consist of pigment cells surrounding microvillous rhabdomeres and they show a degree of bilateral asymmetry in their size. It seems probable that these structures are photosensitive, serving to orientate the larva, and perhaps assisting in locating the snail host. Associated with the eye-spots of at least one larval digenean (*Diplostomum spathaceum*) are ciliated cells whose membranes show marked evaginations. On circumstantial evidence alone, these are thought to be photoreceptors. The functions of the other miracidial sense organs are not known for sure, but they include tangoreception (ciliated nerve endings), chemoreception (ciliated pits and ciliated papillae) and mechano- or rheoreception (lateral papillae).

Cercariae Relatively little is known about the sensory structures in the cercaria, but the presence of eye-spots, papillae and sensory hairs suggest that these larvae are capable of receiving a range of types of information from the environment. The morphology of the eye-spots closely resembles that of the miracidia, i.e. they consist of pigmented cells, rhabdomeres and microvilli. In many strigeoid cercariae, the eye-spots lack pigment, but the functional significance of this is not understood. The sensory papillae of cercariae are thought to be tangoreceptors, but no function has yet been ascribed to the hair-like projections that cover the body surface and which are particularly prevalent on the tail.

Adult digenea   Sensory structures in adult digeneans are associated with the tegument; schistosomes possess ciliated sense organs over the entire body surface, but these are concentrated, in the male worm, in the anterior region of the gynaecophoric canal. Although similar in basic morphology to cercarial sense organs, in the adult, the cilia are completely enclosed by the membranes of the tegument, suggesting that they are mechanoreceptors.

Cestoda

Sense organs associated with the tegument have been described in a small number of species, using standard histological techniques at both light and electron microscope levels. The sense organs are tegumental protrusions, extending from bulbous swellings in the syncytial region. Their function is not known, but it seems probable that adult tapeworms possess chemoreceptors, that provide information on the chemical nature of the environment, and which may allow them to select the optimum region in the host gut for habitation. It has been suggested that proprioreceptors are present in the tapeworm scolex, associated with stretch reception within the organs of attachment. Histological evidence for their occurrence is available only for *Acanthobothrium*.

Acanthocephala

The only information on acanthocephalan sense organs concerns the presence of small pits with a nervous supply, located on the proboscis and around the male copulatory apparatus.

Nematoda

In contrast with the other groups of parasites, there is an extensive literature on the sensory structures of nematodes, and a wide variety of different sense organs have been identified and described (table 9.1). In

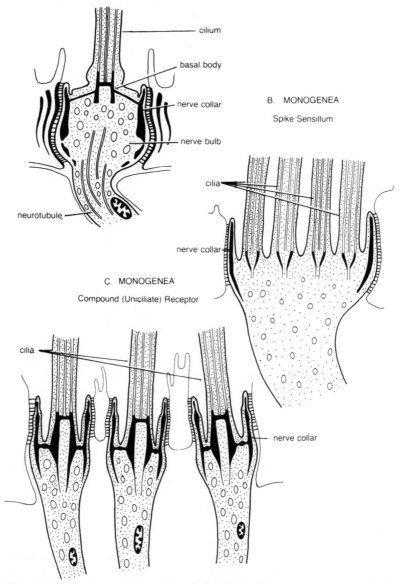

Figure 9.2 Some common sense organs of helminth parasites.
A., B. and C. redrawn from Lyons, K. M. (1972) *Behavioural Aspects of Parasite Transmission.* Linnean Society Monograph, 181–200.

D. DIGENEA   Eye-Spot of Miracidium

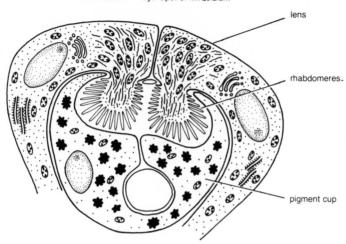

lens

rhabdomeres.

pigment cup

E. DIGENEA   Adult (schistosome) Sense Organ

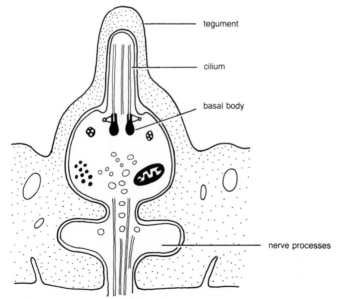

tegument

cilium

basal body

nerve processes

D. redrawn from Isseroff, H. and Cable, R. M. (1968) *Zeitschrift fur Zellforschung*, 86, 511–534.
E. redrawn from Hockley, D. J. (1973) *Advances in Parasitology*, 11, 233–305.

Table 9.1 Nematode sense organs and their functions.

Organ type	Description	Function	Examples
Head			
labial and cephalic papillae	3 concentric circles of papillae around mouth; single cilium, covered by cuticle	mechano- and chemoreception	*Haemonchus, Ascaris, Dipetalonema, Necator, Nippostrongylus, Nematospiroides*
amphids	contain several modified cilia; open by a pore	chemoreception; photoreception (?) associated with secretory gland	many species
cephalids	anterior and posterior glands	unknown	plant parasites
Neck			
deirids	paired, papillate, lateral organs	mechanoreception (?)	*Syngamus, Ascaris*
hemizonid	refractile band, between cuticle and hypodermis	unknown	*Ancylostoma, Trichinella, Haemonchus*
lateral, cervical organs	contains single modified cilium	proprioception (?)	*Ancylostoma, Necator*

bacillary bands	hypodermal cords; sensory cells with gland cells	ionic and osmotic regulation	Trichuroidea
anterior ventral body wall organs	contain 4 cilia; open through pore	chemoreception (?)	Trichuroidea
photoreceptors	usually absent in parasitic forms		*Mermis*
Posterior body and tail			
caudal papillae	around male cloaca	copulatory chemoreception	*Dipetalonema, Syphacea*
spicules	nervous tissue within spicule of male	copulatory mechanoreception	*Heterodera, Heterakis, Nippostrongylus*
phasmids	lateral, paired, with cilia and a pore	sensory and glandular role	*Dracunculus, Necator, Oesophagostomum*
caudalids	hypodermal commissures in anal region	unknown	*Ascaris*
Other sensory structures			
proprioreceptors	intestino-cloacal junction	stretch receptors	*Heterakis*
setae	cuticular bristles over surface		not in parasites

Based on data from McLaren (1976).

general, nematode sense organs are of two types; one with bundles of nerve axons, and the other with nerve endings in the form of modified cilia (amphids, phasmids and papillae). The latter types of sense organs are commonly associated with secretory activities and enzymes, including acetylcholinesterase and other esterases, have been identified, but the sensory role of these organs is not clear.

Nervous transmission, neuromuscular junctions and neurosecretion

The neurophysiology of parasitic animals is an area that has largely been neglected. The probable reasons for this are twofold: firstly, parasites are frequently quoted as being degenerate animals, and hence their sensory physiology would seem to be a poor area of research interest; and, secondly, because of their small size and the difficulties of laboratory maintenance, parasites have proved not to be the best models for neurophysiological investigations.

Acetylcholinesterase activity has been demonstrated in a number of parasites including *Schistosoma mansoni*, *Fasciola hepatica*, *Diplostomum phoxini*, *Diphyllobothrium latum*, *Taenia taeniaeformis*, *Hymenolepis* spp., *Echinococcus granulosus*, *Dipylidium caninum* and many nematodes. Adrenaline stimulates muscle contraction in *Hymenolepis nana*, and both catecholamines and 5-hydroxytryptamine (5-HT) are present in the nervous system of schistosomes. In *S. mansoni* there is pharmacological evidence that acetylcholine inhibits neurotransmission, while 5-hydroxytryptamine acts as a stimulator. It is proposed that catecholamines may act as interneuronal transmitters and stimulate 5-HT release, so transmitting motor nerve impulses to effector organs.

Adult *Ascaris lumbricoides*, being both large and readily obtained from the local slaughterhouse, is a convenient model for studies on neuromuscular physiology. Isolated nerve-muscle preparations have been examined and found to respond to various pharmacological agents rather differently from other animals, e.g. acetylcholine and nicotine stimulate muscle contraction, but adrenaline and histamine have no effect. Excitation of action potentials in the neuro-muscular junction is brought about by acetylcholine and inhibited by γ-aminobutyric acid. Acetylcholinesterase is present in varying amounts in the tissues of *Ascaris* and many nematodes, occurring particularly at nerve-muscle junctions, in many of the sensory structures and in the nerve ring. The muscle cells of nematodes are unique in comprising a contractile portion and a non-contractile portion, which makes contact with the excitatory and inhibitory nervous supply. The muscle cell appears to possess proprio-receptive capabilities, but no receptor has been identified (figure 9.3).

Some of the substances used as commercial anthelminthics affect neuro-muscular transmission in parasites, e.g. bephenium salts, 5-hydroxytryptamine and methyridine either bring about neuromuscular blocking or mimic acetylcholine activity, while piperazine, a commonly used drug against intestinal-dwelling parasites, acts through inducing muscle paralysis. Some of the antischistosome drugs, such as hycanthone, affect parasite motility. More complete information on the neuro-muscular physiology of parasites, in general, will have clear implications in the future development of "logical anthelminthics".

Neurosecretory cells have been identified by histochemical procedures in a small number of parasites, (e.g. *Dicrocoelium dendriticum*, *Hymenolepis diminuta* and *Echinococcus granulosus*, as well as in several species of nematodes). In many nematodes, there is a clear relationship between moulting and neurosecretory activity, and insect juvenile hormones and their synthetic analogues (farnesol and farnesyl methyl ester) can affect the normal moulting process. Bioassay of a substance

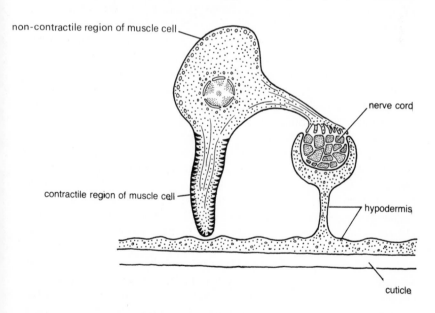

Figure 9.3   The nematode muscle cell. Redrawn from Lee, D. L. and Atkinson, H. J. (1976) *Physiology of Nematodes*, 2nd edition, Macmillan.

isolated from infective larvae of *Haemonchus contortus* reveals insect juvenile hormone activity.

Behavioural coordination

Behavioural studies of helminth parasites are limited to the host-finding activities of larval monogeneans, digeneans and nematodes, and to behavioural analyses of the movements of some adult nematodes. The former has been discussed at some length in chapter 6. Interest in the latter has only recently emerged and is at an early stage of development.

The mechanisms of coordination of movement have been examined in a few nematodes using a variety of techniques, including agar track analysis, cinematography and closed-circuit television studies, and a neurophysiological approach. The processes and mechanics of locomotion in *Ascaris* are known in considerable detail, and the relative roles of the elastic cuticle, the pseudocoelomic hydrostatic skeleton and the body musculature, in producing the characteristic sinusoidal locomotory movements, are now well understood, However, other distinct patterns of movement can be discerned, associated with activities, such as responses to external stimuli, the act of defaecation, host penetration, mating and feeding. It is becoming apparent that parasitic nematodes are complex and behaviourally sophisticated animals in both their coordination of movement and in their responses to environmental stimuli.

The host-finding behaviour of larval monogeneans and digeneans has attracted considerable attention, but the mechanisms involved are not fully understood. Monogenean oncomiracidia swim freely, but slowly (5 mm sec^{-1}) when compared to the speed of the potential host fish to which they must become attached, and which is of the order of 60 to 300 times faster. It seems improbable, therefore, that successful invasion can take place in open waters, so the behaviour of oncomiracidia must be adapted to facilitate invasion when the hosts are congregated (spawning, feeding or shoaling) and rather less mobile than when swimming freely. The reproductive behaviour of the adult parasite must also be timed so that the release of oncomiracidia coincides with the most favourable time for invasion. Strategies associated with this include the infection, by adult monogeneans, of young fishes only (e.g. *Protancyrocephalus strelkovi* on *Limanda aspersa* and *Gastrocotyle trachuri* on *Trachurus trachurus*) or the infection of older fishes exclusively (e.g. *Mazocraes alosa* on *Alosa* spp.). These strategies are related to the age-related spawning activities of the host.

Oncomiracidia, miracidia and cercariae are all short lived stages in the parasite life-cycle, remaining active for usually no more than 24 hours, and maintaining their ability to invade and settle on a host for perhaps considerably shorter periods of time. It is unlikely that infection of a host can take place at a great distance form the site of larval emergence. The responses of parasite larvae to environmental stimuli, such as phototaxis, chemotaxis, geotaxis and rheotaxis, probably serve two functions; one is to place the larva in the most favourable position for making initial contact with the host, while the second serves a short range function concerned with precise location, attachment and, finally, invasion of the host. Unfortunately we have insufficient information on the sensory physiology of larval helminths to make exact statements about their behavioural coordination associated with invasion of the host.

Nematode larvae respond to environmental stimuli, such as light, heat, chemical substances, mechanical events, gravity and electric fields, and their behavioural and locomotory patterns can alter according to the stimuli present in the external environment. Often these behavioural changes are clearly related to host locating activities, e.g. the heat responses shown by larval hookworms, which move rapidly towards a heat source, or the positive phototaxes of pastoral nematodes, such as *Haemonchus contortus*. Many of the other responses recorded seem random and have no apparent advantage, e.g. the responses to changes in electric fields seen in *Heterodera*, *Litomosoides* and *Pelodera*.

Although it is evident that adult helminth parasites respond to environmental stimuli emanating from the host tissues, this aspect of parasite behavioural coordination is poorly understood. Stimuli that bring about site selection and diurnal rhythmicity, and the ways in which parasites receive, and respond to, this information are yet to be determined.

SUMMARY

1. All parasitic helminths have a relatively simple nervous system, the morphology of which is described.

2. Surface sense organs of parasites consist of pigmented and non-pigmented eye-spots, ciliated photoreceptors, papillae, ciliated pits and ciliated protruberances. The function of these sense organs is deduced exclusively from studies on their morphology. Parasitic nematodes possess a considerable range of sensory structures.

3. The neurophysiology of parasites is not well known. It is an area of particular importance in the chemotherapy of various parasitic diseases.

4. Study of the behavioural coordination of parasites, although in its infancy, reveals a complexity of pattern. Behavioural responses associated with location of the host by the parasite have attracted the greatest interest.

CHAPTER TEN

HOST-PARASITE INTERACTIONS

Introduction

All organisms have evolved defence mechanisms for protection against invasion and establishment by micro-organisms and by metazoan parasites. Although ideally we should include the viruses and bacteria in our current discussion, space precludes such a full consideration. Before discussing immunity to parasites, it will be necessary to outline, in the briefest manner, the components of the defence systems—primarily of mammals.

The discipline of immunology has developed enormously over the last few decades, and it is often difficult to keep abreast of new developments. Although our knowledge of immunity to parasites is perhaps less complete than that in the areas of tumour and micro-organism immunology, a vast new literature is accumulating on the highly varied immune defence mechanisms adopted by mammals against parasitic infection. Nevertheless, many current textbooks of immunology pay little or no attention to parasite immunology, which perhaps reflects more the difficulties embodied in the subject, rather than the paucity of information. The immunology of parasitic infections is both an important academic subject, as well as a fundamental applied discipline. It is to the field of immunological achievement that many people from the Third World will be looking for relief from the major parasitic diseases that reduce the quality or length of their lives.

General principles of cellular and immunological defence systems

The essence of the immunological defence system of vertebrates is the recognition of self and the ability to distinguish it from non-self, i.e. the host animal must be able to distinguish between its own macro-molecules and those of an invading organism. On occasions this system of recognition can fail, giving rise to the so called auto-immune diseases, in which the host's defences act disastrously against its own tissue. Nevertheless, in the normal healthy animal, the ability to recognize non-self material and the development of a specific memory system associated with that recognition is the embodiment of the immune system.

Defence mechanisms of vertebrates against invading organisms are divisible into either specific or non-specific responses. Specific responses are associated with an immunological memory component and are normally directed against secondary and subsequent infections of an organism, whose primary infection initiates the response. Therefore, once a host has been infected with a particular parasite it retains an immunological memory of that infection and responds to eliminate or control any later infections by that parasite. Non-specific responses are more generalized in their nature, and are directed against any invading organism, with no involvement of a memory element. In some cases, specific and non-specific responses cooperate to form a combined defence against invasion.

Non-specific responses tend to have a cellular basis and involve the action of phagocytic cells (macrophages and polymorphonuclear leucocytes), which form the reticuloendothelial system and are present in the circulating blood and in the tissues of the host. Their function is to engulf invading organisms and digest them lysosomally. Engulfment and digestion are sequential processes: first, the invading organism becomes adhered to the surface of the phagoctyic cell; secondly, it is ingested by the formation of a vacuole, becoming trapped within a phagosome; thirdly, a lysosome fuses with the phagosome to form a phagolysosome; and finally, the engulfed organism is destroyed by an array of lytic factors, including enzymes. Phagocytosis is a common mode of defence against invading micro-organisms, and it can be enhanced by the action of antibodies known specifically as *opsonins*. The complex of serum factors, known as *complement*, may also aid the process of phagocytosis by attracting phagocytic cells to the environs of an invading parasite. Other non-specific responses to parasites may take the form of inflammatory reactions and capsule formation. Inflammation is typified by increased capillary blood supply to a site of invasion, resulting in the accumulation of polymorphs around the invading parasite. Some polymorphs can transform to fibroblasts, which contribute to encapsulation and ultimately, calcification of the parasite; digenean metacercariae and pseudophyllidean plerocercoids are often encapsulated in this way in their intermediate, fish hosts.

Specific immune responses involve the formation of antibodies by the host as a result of stimulation by the presence of foreign material, termed antigens. Antigenic substances are usually polypeptides, proteins or polysaccharides, with a molecular weight in excess of 5 000 Daltons (lipids are only antigenic when combined with other substances, called haptens). Antigens derive from the surface coat of a parasite, i.e. surface antigens, or from secreted or metabolic byproducts, i.e. soluble or metabolic antigens. Substances are only antigenic if they exhibit a degree of foreigness to the host, which then responds by synthesizing antibodies, specific to each antigen.

Antibodies are soluble serum proteins that combine, specifically, with antigens. They belong to the gamma-globulin fraction of serum proteins (figure 10.1a), and are referred to as *immunoglobulins* (Ig). They are separable into five distinct classes, depending upon their structure, i.e. IgA, IgD, IgE, IgG and IgM (table 10.1). The molecular structure of the immunoglobulins consists of two heavy peptide chains, linked by disulphide bridges, and two light chains, that are also cross-linked. Papain treatment cleaves the immunoglobulin molecule into two fractions, the F*ab* (antigen binding fragment) and the F*c* (crystallisable fragment) components (figure 10.1b). Antigen binding takes place at the F*ab* end of the immunoglobulin molecule, and thus, antibody function is retained after papain cleavage of the intact molecule.

Immune defence effected by antibodies is divisible into two distinct processes, known as *cell-mediated immunity* (delayed hypersensitivity) and *humoral immunity*. Both types of immunity arise from stem cells produced in bone marrow. In cell-mediated immunity, the stem cells are processed by the thymus gland to form a population of lymphocytes (T-lymphocytes) which become sensitized upon contact with antigen, in the paracortical region of lymph nodes. These T-lymphocytes are long-lived and carry firmly bound antibody on their surface membranes. In humoral immunity, the bone marrow stem cells are processed in lymphoid tissue (Bursa Fabricius in birds and Peyers Patches in mammals) to give rise to a population of B-lymphocytes, which are short-lived, after contact with antigen. These lymphocytes differentiate into plasma cells which synthesize and secrete antibody into the surrounding blood and tissues (figure 10.2). Cellular co-operation may enhance an immune response, and such co-operation can involve contact between macrophages and lymphocytes or between B-cells and T-cells (helper T-cells). Cell mediated immunity is involved primarily in allergic responses and can take some time to develop (24 hours or longer), whereas humoral immunity is developed much more rapidly.

A. ELECTROPHORETIC SEPARATION OF SERUM PROTEINS

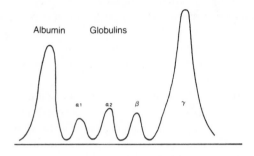

B. THE STRUCTURE OF THE IMMUNOGLOBULINS (ANTIBODIES)

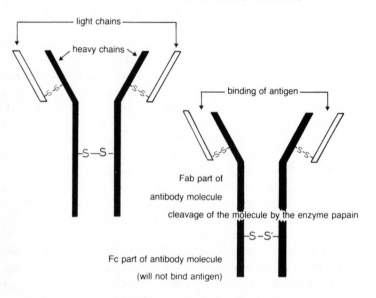

Figure 10.1 The identity and structure of mammalian immunoglobulins (antibodies). Based on Roitt, I. (1977) *Essential Immunology*. Blackwell.

Table 10.1 Major properties and functions of the immunoglobulin classes.

Property	IgG	IgA	IgM	IgD	IgE
molecular weight	1.5×10^5	1.6×10^5	9×10^5	1.85×10^5	2×10^5
valency of antigen binding	2	2	5 (10)	?	2
serum concentration (mg/ml)	8–16	1.4–4	0.5–2	0–0.4	0.017–0.45
carbohydrate content (%)	3	8	12	13	12
major known role	anti-micro-organisms	defence of body surfaces	agglutination: primary immune defence	found on lymphocyte surfaces	anti-parasite action: reaginic antibody (allergic reactions).

Data from Roitt (1977)

Antibodies react with antigen and effect lysis, agglutination, precipitation or enhanced phagocytosis of the cells producing the antigens. These reactions can form the basis of *in vitro* tests used to detect the presence of specific antibodies to a parasite, and they serve to determine whether an individual animal has been exposed to a particular parasite (immuno- or sero-diagnosis).

Immunological defence is best developed and most often studied in mammals, and rather less is known for lower vertebrates and invertebrates. In non-mammalian vertebrates, humoral and cell-mediated immunity can be identified, but invertebrates appear to lack antibodies, in the mammalian sense, and rely, to a great extent on, phagocytosis as a defence against invasion.

Passive immunity is passed from a mother to her offspring and reflects the innate resistance to a particular infection. Passive immunity can also

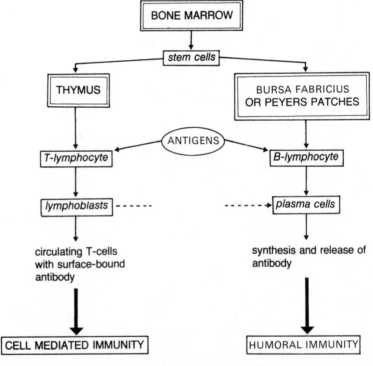

Figure 10.2 Lymphocytes and antibody production: A simplified scheme.
Based on information from Roitt, I. (1977), *Essential Immunology*, Blackwell.

be transmitted in the laboratory by the donation of serum or immunologically competent cells to recipient, naive animals. *Acquired immunity* depends upon the experience of a primary infection to initiate the specific immunological processes that render the animal insusceptible, or less susceptible, to a second or subsequent infection.

Immunity to Protozoa

Blood-dwelling Protozoa

In this section we shall examine briefly the immunity to malaria, babesias and trypanosomes, which together constitute the major blood-dwelling parasitic protozoans of vertebrates.

Malaria The general picture of immunity to malarial parasites is remarkably complex due to a range of factors that include parasite strain variation, antigenic variation within strains, and a life-cycle that involves intraerythrocytic stages and exoerythrocytic stages, both in the blood and in liver tissue. Acquired immunity to *Plasmodium* is variable, e.g. premunition is typical of avian malaria, whereas sterile immunity occurs in some rodent malarias, such as *Plasmodium berghei* and *P. vinckei*. In the former case, parasites persist in spite of the development of immunity, while in the latter case immunity is complete and no parasite survives. Premunition is thought to typify human malarial infections, in which relapses and recrudescences are not uncommon.

Protective immunity to malaria is recognized as a collaborative mechanism involving both cell-mediated reactions (T-lymphocytes) and humoral antibodies (B-lymphocytes). Immunoglobulin production increases considerably during the course of a malarial infection, but much of the antibody produced is not protective and does not serve to eliminate the parasite. Recovery from malaria requires protective antibody to be manufactured at a rate sufficient to control the multiplying parasite; drug induced immunosuppression in experimental malarias, can often result in a fatal infection developing from what would otherwise have been a non-lethal, self-limiting condition. There is considerable variation in the levels of the various immunoglobin classes, e.g. IgM levels are typically high during the early stages of an infection, while IgG levels rise slowly and are maintained for much longer periods. The roles of IgA, IgD and IgE are not clear in protection against malaria, and this function appears to reside primarily with IgG. This is supported by the observation that, during pregnancy, IgG levels characteristically decrease, and there is often an enhancement of malaria in pregnancy.

Cell-mediated reactions, involving lymphokine production, the potentiation of antibody production from B-lymphocytes by helper T-cells and cytotoxicity, may all be involved in the development of immunity to malarias, but current knowledge remains incomplete. Surgical removal of the thymus (thymectomy) in neonatal rats and hamsters, renders them more susceptible to *P. berghei* infections than intact controls. Depression of T-cell function by the administration of anti-T-cell serum, can also exacerbate malaria in experimental animals.

There is a number of diagnostic tools, based on immunoglobulin production in response to malarial infection, which serve to identify individuals infected with the parasite, where microscopic examination might prove negative. Such immunodiagnoses include the indirect fluorescent antibody test, the indirect haemagglutination test and various gel precipitation tests. The application of these methods to human populations can provide considerable and invaluable information on the epidemiology of malaria in endemic regions of the world. Serological monitoring is also proving to be an important tool in the prevention of transfusion-induced malaria, which is an increasing problem related to the degree and ease of world travel that typifies the late twentieth century.

Malarial parasites can avoid the immune responses of their hosts to form residual populations, either in red blood cells or within liver tissue, and these residual populations of parasites are able to initiate recurrent relapses or recrudescences. The latter may result either from infected red cells that are situated at deep vascular sites, the plasmodia re-emerging as the effective immunity weakens, or from either the continued existence of merozoites in the circulating blood or sporozoites in the liver. Antigenic variation, whereby a sub-population of parasites, of different antigenic structure from the original invasion, is developed, may also be important in this context. This new antigenic population can now withstand the effects of the antibodies directed against the antigens of the initial infection. In mammalian malaria, antigenic variation represents an important problem, hindering the development of a vaccine.

The pathology of mammalian malaria is related, in part, to the immunological responses of the host to the parasite—hence the term immunopathology. Red cell destruction (haemolysis) and enlargement of the spleen (splenomegaly) are typical of advanced malaria. Haemolysis may be brought about by the phagocytic activities of spleen macrophages, and both infected and uninfected red cells are destroyed. Opsonising antibodies enhance this pathological process. Haemolysis in Blackwater fever (*P. falciparum* infections associated with quinine therapy) occurs in the vascular system and is not related to the splenic mechanism. Malarial kidney disease results from antigen-antibody complexes binding to the glomerular basement membrane, producing yet another aspect of malarial immunopathology.

An interesting feature of malarial parasites lies in their role as immunosuppressive agents for infections by other parasites. Laboratory animals, infected with various species of *Plasmodium*, show a reduced immune response to non-living antigen, such as tetanus toxoid or sheep red blood cells, and to living antigens in the form of viruses, bacteria and several protozoans. It is not known whether this immunosuppressive effect is functionally related to cell-mediated or humoral immunity. There is good evidence for suppression of antibody production by B-lymphocytes, but the general mechanisms by which immuno-suppression to unrelated antigens is effected remain to be elucidated.

Toxin production in malaria has been a much debated issue for many years. The key question of whether the parasite releases a toxic substance, which itself is pathogenic, has yet to be answered. There is some evidence that a transferable serum component will induce haemolysis in recipient animals, and that serum elements are capable of inhibiting mitochondrial respiration and oxidative phosphorylation in liver cells *in vitro*. Antitoxic immunity, however, has no effect on the levels of parasites in the blood, but may serve to alleviate aspects of the immunopathogenesis.

Babesia Members of the genus *Babesia* are intraerythrocytic parasites of vertebrates, occurring most commonly in mammals. Several species are of veterinary importance as they cause a variety of tick-borne fevers of domestic animals, e.g. *B. bovis*, red-water fever; *B. equi*, biliary fever. Immunity to many species of *Babesia* is of the premune type, existing only while the host remains infected by the parasite. For some species, however, complete sterile immunity is developed, e.g. *B. divergens*, *B. bigemina* and *B. rhodaini*. Immunity may be highly specific for strains of a single species, such that cattle, immune to a strain from one locality, can be susceptible to a different strain of the same species from other localities. In other, cases cross-immunity, between different species, is found, e.g. *B. rhodaini* and *B. microti* in rodents. The extent of antigenic variation in babesias is not known.

Trypanosomes   Extensive investigations have been made into immunity to both rodent and human trypanosomes. One of the best known model systems is *Trypanosoma lewisi*, a specific rat parasite, which provides a classical demonstration of acquired immunity. Additional interest in *T. lewisi* is generated by the occurrence of an apparently unique antibody, termed ablastin, whose functional significance has proved to be controversial. Typically, the course of a *T. lewisi* infection in a laboratory rat is characterized by an initial phase of rapid multiplication of the parasite, giving a rising parasitaemia. Parasite division is inhibited five days post-infection and a plateau in the numbers of trypanosomes occurs, but this is of short duration. Subsequently, a primary crisis takes place on or about the tenth day of the infection, during which most of the trypanosomes are killed. The few parasites that remain survive until approximately the thirty-fifth day, when a second crisis completely eliminates the infection (figure 10.3). One explanation of these events involves the production of three distinct antibodies, one that inhibits reproduction of the parasite (ablastin) and two lethal (trypanocidal) antibodies, which contribute to the two crises.

The name ablastin was coined by Taliaferro, more than fifty years ago, for the antibody that inhibited the reproduction of *T. lewisi* but lacked

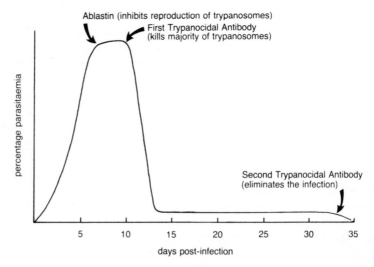

Figure 10.3   The course of an infection of *Trypanosoma lewisi* in a rat: one possible scheme of antibody action in eliminating the infection. Alternative hypotheses to account for the course of the infection are discussed briefly in the text.

trypanocidal activity. Since the original experiments which demonstrated ablastic immunity were completed an interesting and controversial literature has arisen. Ablastin is found only in infection of two species of trypanosomes, *T. lewisi* and *T. musculi*, although ablastic-like responses are characteristic of many stercorarian trypanosome infections. Despite intensive investigation, ablastin has not been unequivocally identified as an immunoglobulin, but it does possess physico-chemical properties typical of IgG. However, ablastin is not adsorbed from infected rat serum by trypanosomes *in vitro*, a feature which throws some doubt on its role as an antibody. Furthermore, the trypanosome antigen (ablastinogen) that induces ablastin production, has yet to be identified, though there is some evidence that this antigen is released into the blood of the host during the course of a normal infection. Pregnant female mice are unable to limit infections of *T. musculi*, suggesting that ablastin belongs to the IgG class of immunoglobulins, which are known to decrease during pregnancy. Some authors, however, regard the apparent uniqueness of ablastin as sufficient reason, in itself, for discounting the presence of an antibody that is not lethal but inhibits only parasite reproduction. One alternative explanation that has been advanced, invokes the occurrence of two parasite antigens, one juvenile and the other adult, that stimulate the host to manufacture trypanocidal antibodies. Antibody against juvenile antigen eliminates juvenile trypanosomes by the tenth day post-infection, while adult parasites survive for longer periods, being— according to this hypothesis—less antigenic. Additionally, it has been proposed that ablastin may act as a metamorphosis-conversion-factor in the development of trypanosomes during an infection, enhancing the change from juvenile to adult form. The first crisis in an infection would then be attributable to the activities of a trypanocidal, anti-juvenile antibody, which may even be ablastin. These views contrast with the classical view of immunity to *T. lewisi* as depicted in figure 10.3. Despite considerable speculation, the nature of a ablastin and ablastic immunity remains unresolved. Its apparent uniqueness as an antibody argues for generating research interest rather than for a denial of its unusual role.

Studies on immunity to the human trypanosomes, causing sleeping sickness, e.g. *Trypanosoma gambiense* and *T. rhodesiense*, are complicated by the appearance of *antigenic variation*, which typifies these haemo-flagellates. Each species has a highly complicated antigenic structure, which gives rise to both strain and variant populations. These render attempts to immunize humans or domestic cattle against trypanosome infections almost futile. The earliest observations on antigenic variation were made at the beginning of this century, when it was noted that *T. equinum*, isolated from the blood of a monkey, was unaffected by serum antibodies from the same animal *in vitro*, and it was concluded that the trypanosomes had undergone antigenic change during the course of the infection. Similar observations to this have been made on many subsequent occasions. Antigenic variation can give rise to a *cyclic parasitaemia*, in which the numbers of parasites increase until a crisis occurs which eliminates the majority of individuals, except for a small number of antigenically variant forms which survive and divide to produce a second, rising parasitaemia, followed, in time, by a second crisis. This pattern of increasing and decreasing parasitaemia may continue until the host dies (figure 10.4). Experimental observations show the frequency with which new antigenic variants can arise, e.g. *T. congolense* produces a new variant every ninth day in sheep, and *T. vivax* develops new variants every seven days in sheep and calves. The maximum number of different variants arising from a single strain is 23,

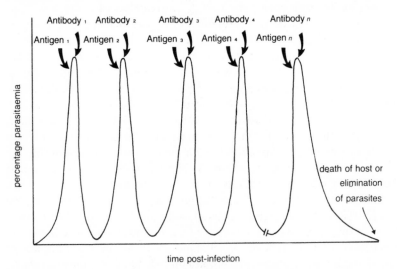

Figure 10.4 Antigenic variation and relapsing infections in trypanosomes. Antigens 1, 2, 3, 4, . . . n are variants produced by a population of trypanosomes in a single host. Each variant induces the production of a variant-specific antibody which eliminates the majority of the trypanosomes. A small number of parasites survive because they possess the new variant against which the host has not made antibody and they multiply to give a rising parasitaemia until there is sufficient variant-specific antibody circulating to eliminate this sub-population. The cyclic process can continue until the death of the host or the complete elimination of the parasites. This explains the characteristic relapsing nature of human sleeping sickness.

but this is thought unlikely to represent the upper limit. Only rarely will the same variant appear twice during a single infection. There is also evidence that antigenic variation is an organised process, with a particular strain of trypanosomes producing the same variants, often in a similar sequence, regardless of host species.

Two major subclasses of antigens have been identified from trypanosomes. The first are the bound antigens which remain stable during a relapsing infection. These antigens include enzymes and nucleoproteins and are only weakly antigenic. The second subclass contains the surface-coat antigens, which are unstable entities. The surface-coat antigens are the glycoprotein exoantigens that are present in the serum of infected animals and are responsible for inducing the production of protective antibodies.

Antigenic variation in trypanosomes may be regarded as an adaptive mechanism to overcome the immunological defences of the host. Its occurrence makes it most unlikely that successful immunisation will be accomplished in the forseeable future.

Tissue-dwelling Protozoa

Among the major tissue-dwelling protozoans of importance to man from a medical or veterinary point of view are the leishmanias, the coccidians and the amoebae.

Leishmania Various species of the genus *Leishmania* cause very unpleasant and often lethal diseases of man in many regions of the world. These include oriental sore (*L. tropica*), kala azar (*L. donovani*) and New World cutaneous leishmaniasis (*L. mexicana* and *L. braziliensis*). The parasite is transmitted to man via the bite of infected phlebotomine flies (sand-flies) and, after entry, the amistigote form of the parasite invades and inhabits the cells (such as monocytes and other phagocytic cells) of the reticuloendothelial system. Immunity to *L. tropica* develops after an initial infection, i.e. acquired immunity. The parasite invades the skin macrophages and a superficial sore develops. This sore will heal spontaneously, after which immunity to reinfection is complete. Should the sore be removed prematurely by surgery, however, the individual remains susceptible to reinfection. Circulating antibodies cannot be demonstrated in *L. tropica* infections, whereas humoral components are readily observed in *L. donovani* infections. In general, it is considered that the human immunological response to leishmaniasis is primarily via cell-mediated immunity, with circulating antibodies playing only a minor protective role. A population of specifically sensitized, thymus-dependent lymphocytes develops on contact with parasite antigens, and these cells destroy the infected macrophages. The actual mechanism by which the parasite is eliminated is not fully understood, but it may occur by a number of processes, including phagocytosis by activated macrophages, cytoxic effects of lymphokines secreted by sensitized lymphocytes at the infected macrophage surface, or, perhaps, by lytic antibody activity.

Some degree of cross-immunity occurs in leishmaniasis, e.g. *L. tropica* protects an individual against infection by *L. mexicana* and *L. braziliensis*, but not against *L. donovani*.

Coccidia Most domestic animals appear to develop immunity to reinfection by species of the genus *Eimeria*, which are obligatory, intracellular, gut parasites. Sera of infected animals effectively prevents infection by the invasive stages of *E. tenella*, *E. bovis* and *E. meleagrimitis*, and the presence of circulating antibodies has been demonstrated by a variety of techniques. Rather surprisingly, passive transfer of immunity, by the injection of immune serum into uninfected hosts, has not proved to be particularly successful, and is only effective if the challenge infection is administered shortly after the inoculation of immune serum from the donor animal. Immunity to *Eimeria* is highly species-specific, so that there is no cross-immunity between different species. The precise mechanisms of acquired immunity to *Eimeria*, either in chickens or laboratory rodents, is not clear. Humoral antibodies appear to be implicated in the development of protective immunity, while there is some evidence that cellular elements, polymorphonuclear cells, lymphocytes and macrophages, are also involved.

Protective antinodies are produced against another important intra-cellular gut parasite, *Toxoplasma gondii*, but it is thought that the cystic stage of the parasite is resistant to these antibodies.

Amoebae *Entamoeba histolytica* is normally a non-pathogenic in-habitant of the human intestine and, under such conditions, is apparently non-immunogenic. However, the amoebae can occasionally penetrate the gut epithelium and spread throughout the body tissues. When this occurs antibodies are synthesized by the host, and serological tests can be used to identify their presence and thus serve to diagnose amoebiasis. At least two antibody systems have been identified, but it is not certain by which mechanism protective immunity is developed.

Immunity to helminths

Metazoan parasites differ fundamentally from protozoans in that the great majority do not multiply within the final host, unlike the Protozoa. Thus immunity to helminths will be directed primarily against the individual parasite, rather than its reproductive ability (Cf. *Trypanosoma lewisi*). There is increasing interest in the immunology of mammalian helminth infections, particularly with regard to those species that affect man's health or his economy.

While protozoan infections are frequently accompanied by an increase in the serum levels of IgG and IgM, helminth infections typically induce the synthesis of reaginic antibody, IgE, which is characteristically abundant in many helminthiases.

Digenea

Information on the immunology of digenean infections depends largely on studies of only two groups, the schistosomes and the liver flukes, e.g. *Fasciola hepatica*.

There is some doubt as to the occurrence of acquired immunity to *Fasciola* in cattle and sheep. Cell-mediated reactions may play an important role in the development of protective immunity in domestic animals. In the laboratory, experimental mice treated with peritoneal exudate cells from infected donors support fewer worms than control animals. T-cell responses may also be responsible for effecting repair of the tissue damage wrought by the parasite, since rabbits given antilymphocyte serum, to suppress T-cell responses, die of liver disease due to the parasite. Transfer of lymph node and spleen cells from infected donor rats to naive recipient rats offers a significant degree of protection against a primary challenge infection, but this is not repeatable using sheep.

The immunology of schistosome infections, particularly *S. mansoni*, has been examined in considerable detail using model systems in rats, mice and rhesus monkeys. Experimental studies in man are, naturally, not acceptable, so that evidence for acquired immunity in human infections derives mainly from epidemiological sources. In general, the young are more susceptible to infection than adult humans, and a premune condition develops, in which a primary infection renders the host immune to subsequent reinfection. In schistosomiasis, this condition is termed *concomitant immunity*, since the primary infection is not killed by the immune response. This situation has led to the suggestion that it may be possible to immunize against infection using attenuated (non-developing) cercariae. There are, however, problems associated with this approach, so the search continues for a dead vaccine that will have fewer attendant risks and be more easily administered.

In experimental infections, acquired immunity of the concomitant type has been clearly demonstrated. Despite the development of a lethal immune response, the primary infection continues to thrive and produce eggs for considerable lengths of time. In rhesus monkeys, immunity to a challenge infection can be developed within sixteen weeks of adminis-tration of a primary burden of 100 cercariae per animal. Mice take much longer to develop resistance to a challenge infection. A *self-cure* type response has been recorded in heavily infected baboons, in which the primary infection is completely eliminated; this does not accompany light infection doses.

The process of concomitant immunity has puzzled researchers for more than a decade, and several explanations for the observed effects have been put forward. It is generally accepted that concomitant immunity arises through the sharing of common antigens by host and parasite, but it is not clear whether the parasite synthesizes host-like antigens or merely binds antigenic material of host origin to its surface-coat. Support for the latter hypothesis comes from ingenious experiments in which schistosomes are grown to maturity in mice and then transferred to the hepatic portal system of rhesus monkeys that have been immunized against mouse red blood cells. Worms transferred in this way are killed within 24 hours, whereas mouse worms will live in non-immunized monkeys. Similar results have been obtained using worms grown, *in vitro*, in human serum and then transferred to monkeys immunized against human erythrocytes. These results support the notion that schistosomes bind host antigens to their surfaces, and by doing so become essentially disguised, rendering them indistinguishable from the host as far as the immune recognition system is concerned. Challenge infections are thought not to be able to bind host antigen at a rate sufficient to elude the antibodies produced against the primary infection. Lethal antibodies have been demonstrated in immune monkey serum, killing schistosomula in culture within three to four days.

Several antibodies have been isolated from schistosome-infected human and animal serum (many of which are used as sero-diagnostic tools), but it seems unlikely that any are directly involved in acquired immunity. Thus, while schistosomiasis is characterized by an increase in the serum levels of IgG, much of this is non-specific. The general inability to demonstrate the passive transfer of resistance to infection with immune serum has led some to believe that cell-mediated reactions are responsible for acquired immunity in schistosomiasis. Recent experiments on passive transfer have proved more successful when rats and mice are used, but not so with monkeys. Additionally the occurrence of a lethal serum component, capable of killing schistosomula *in vitro*, strongly suggests that humoral antibodies are, indeed, required for acquired immunity. It may well be that both B-cells and T-cells, working in a cooperative manner, will prove to be the elements involved, but further evidence is required to establish this hypothesis.

There is good evidence that cell-mediated reactions are concerned with the immunopathology of schistosomiasis. The disease, as seen in the model system of *S. mansoni* in the laboratory mouse, is characterized by liver and spleen enlargement (hepatomegaly and splenomegaly), by portal hypertension (increased blood pressure in the hepatic portal vein) and by oesophageal varices. Egg production by the worms is a fundamental feature in disease production. Typical hepatosplenic schistosomiasis does not develop in either unisexual infections of mice or in normal, bisexual infections of mice given drugs to inhibit the production of eggs by the parasites. The formation of granulomata (nodules of inflamed tissue) around eggs has been studied, experimentally, by examining pulmonary inflammation around schistosome eggs that are administered to mice via the tail vein and which become lodged in the lungs. These studies implicate cell-mediated reactions in ganuloma formation. Immunosuppressant drugs will inhibit the production of granulomata around eggs, but there is no evidence to suggest that humoral antibodies are involved in the inflammatory response to the eggs of either *S. mansoni* or *S. haematobium*.

Partially successful attempts have been made to isolate the egg antigens responsible for initiating granuloma formation, which is thought to be due to the secretion of soluble egg antigens through pores in the egg shell. It is probable that soluble egg antigen stimulates the production and secretion of lymphokine by T-lymphocytes, and this attracts mononuclear cells to the egg. Soluble egg antigen includes the enzymes, secreted by the eggs to aid their passage through the host tissues, on their way to the external environment. In contrast with *S. mansoni* and *S. haematobium*, granuloma formation in *S. japonicum* is mediated by antigen-antibody reactions.

Cestoda

Tapeworms live as adults in the lumen of the vertebrate intestine and as larvae in the tissues of either invertebrates or vertebrates. The immunological status of lumen-dwelling cestodes tends to be somewhat different from that of the tissue stages, since the latter are in much more intimate contact with the host's defence mechanisms than the former.

Acquired immunity to tissue-dwelling larval cestodes operates either at the transition phase, during which the oncosphere hatches in the intestine of the host and invades the mucosa, or during the actual tissue phase itself. Liberated oncospheres are susceptible to immunological attack in the host gut, as shown by studies on *Taenia pisiformis*, *Hymenolepis nana* and *Echinococcus granulosus*. In immune hosts, the oncospheres of these worms either fail to hatch, or hatch but fail to penetrate the gut wall. Immune serum is capable of damaging the oncospheres of *T. pisiformis, in vitro*. Rats harbouring the cysts of *T. taeniaeformis* in their livers are resistant to challenge infections for several months, and a similar pattern of immunity occurs in other larval taeniids, e.g. *T. hydatigena, T. ovis, T. saginata* and *T. pisiformis*. Passive transfer of immunity is accomplished by injecting whole immune serum or lymph node cells from immune hosts. Thymectomy reduces protective immunity of mice to *H. nana*, a tapeworm whose cysticercoids can develop in the intestinal mucosa of the definitive host. Protection against reinfection is restored by implantation of thymus tissue. In *H. nana* infections, passive protection is also associated with serum antibodies, particularly IgG. Studies on the hydatid stage of *Echinococcus granulosus* suggest that a complete protective immunity does not develop, since challenge infections are numerically less successful than the primary, immunizing infection, but they still survive.

Although contact with host tissues tends to be less intimate in many of the lumen-dwelling adult cestodes, they are, nevertheless, immunogenic. Current interest has focussed on *Hymenolepis diminuta, H. microstoma, Raillietina cesticillus* and *Echinococcus granulosus*. *Hymenolepis diminuta* develops very differently when grown in laboratory rats and mice. In the mouse, this parasite grows normally for ten days, then growth ceases and destrobilation occurs. Transfer of destrobilated worms to rats restores normal growth. A challenge infection given to immune mice (infected and cleared of worms by anthelminthic treatment) dies more rapidly than a primary infection. Additional evidence for the immunological basis of these events derives from the use of immunosuppressants, e.g.

cortisone, methotrexate and antilymphocyte serum prevent both the rejection response and destrobilation. The rat, on the other hand, is the natural host for *H. diminuta* and will support low density infections, i.e. containing between one and ten worms, indefinitely. Destrobilation and worm loss is a feature of higher density infections, but it is not clear whether this is a function of the immune response of the host or if it is due to physiological alterations associated with crowding of worms in the intestine. Recent evidence supports the former proposition; challenge infections of 50 or 100 worms survive less well than primary infections, an effect alleviated by cortisone treatment. Antibodies of various classes are bound to the tapeworm surface in the rat gut and it may be that this binding interferes with the absorption of essential nutrients. *Hymenolepis microstoma*, which normally develops in the bile duct of mice, fails to grow or survive in the rat, but it has not been established that this is due to the immune response of the host. In the mouse, *H. microstoma* infections are accompanied by increased levels of circulating antibodies (IgA, IgE and IgG), but these do not appear to play a protective role.

In general, the early developmental stages of cestodes are more immunogenic than mature worms, so that the intermediate vertebrate host develops a strong protective immunity. Acquired immunity to lumen-dwelling adult cestodes may play a controlling rather than an eliminating role. Larval cestodes that develop in invertebrate animals are exposed to the haemocytic defences, which are either phagocytic or encapsulating processes. There is some evidence that tapeworm larvae, living in the haemocoel of insects, can either evade this defence or disguise themselves so that the host haemocytes fail to identify them as foreign, e.g. *H. diminuta* in flour beetles.

Nematoda

Interest in the immunity of mammals to nematode parasites is well established, and has focussed particularly on the pathogenic species of economic or medical importance to man, or on convenient laboratory systems. Nematodes differ from other parasites in the presentation to the host of a cuticular surface, which is only weakly immunogenic. Furthermore, it is unlikely that antibodies or cellular defences will be able to exert any lethal effect by attaching to the nematode cuticle, and only the anal, oral and glandular openings of the parasite will be vulnerable.

Most nematode infections, and indeed helminth infections in general, stimulate the host to produce reaginic antibodies (IgE), but these antibodies do not appear to be exclusively responsible for limiting an infection. The parasite antigens that stimulate IgE production are termed *allergens*, and these have been identified in some nematodes. There is a relationship between IgE production, eosinophilia and immuno-pathology in many nematode infections, and this may be mediated via anaphylaxis, involving the synthesis and secretion of histamine from mast cells at the site of invasion or habitation of the parasite. The phenomenon called self-cure is probably related to the release of histamine in the gut of hosts infected with lumen-dwelling nematodes, and this causes leakage of plasma proteins into the gut, bringing about expulsion of the worms. Self-cure occurs in *Haemonchus contortus* infections of lambs and *Nippostrongylus brasiliensis* infections of rats, and may prove to be a common expulsive mechanism for gut-dwelling nematodes.

Haemonchus contortus is an economically important parasite, causing pronounced anaemia and often killing young lambs. Self-cure of haemonchosis was first described almost 50 years ago. It is characterized by a sudden decrease in the numbers of parasite eggs present in the faeces of lambs between 12 and 13 weeks post-infection. This immunity is precipitated by the ingestion of additional larval nematodes from the pasture and occurs in lambs over four months old. Vaccination against *Haemonchus* is partially effective using irradiated third-stage larvae, that can migrate but do not develop to maturity. Variation in response by different breeds of sheep diminishes the usefulness of this approach to artificial protection. It has been proposed that the antigens responsible for stimulating self-cure arise from the exsheathing fluid of the third moult.

Self-cure is typical of *Nippostrongylus brasiliensis* infections in the laboratory rat. In heavily infected rats, parasite egg production rises between six and ten days post-infection and then decreases rapidly, accompanied by the expulsion of the majority of the worms. The rats are now highly resistant to a challenge infection. By contrast, worms given to rats as repeated low doses, called *trickle infections*, fail to stimulate any protective immunity and these worms will survive and produce eggs for extended periods, being described as "adapted" and capable of surviving transfer to an immune host. Self-cure in *Nippostrongylus* was originally thought to be due to two distinct events; antibodies inhibiting worm acetylcholinesterase activity and immunopathological effects upon the rat gut epithelium rendering it an unsuitable environment for the worms.

The latter is related to IgE production, stimulating histamine release from mast cells. Other immunoglobulins, including IgA and IgG, are thought to play some role in self-cure. It is apparent, on examining the immunological response of the rat to heavy infections of *Nippostrongylus*, that several aspects of the immune system must operate coincidentally to effect worm expulsion. Both antibodies and sensitized lymphocytes are required for the rejection process, whereas IgE, mast cells and eosinophils are now regarded as non-essential components. Protective antibodies belong to the IgG class of immunoglobulins but not to IgA. The former have been shown to interfere with worm metabolism and cause damage to the worm's alimentary tract.

Trichinella spiralis, causing trichinosis in man, is a good model parasite for the study of acquired immunity using mice, rats and guinea pigs as laboratory hosts. Trichinosis in man is not accompanied by a noticeable increase in serum antibody levels, but repeated infections of experimental animals results in a greatly increased immunoglobulin (precipitin) titre. In rabbits, IgM levels rise early in an infection, followed later by elevated levels of IgG. IgA remains consistently high throughout an infection, and forms a localized mucosal response to the parasite. Trichinosis is characterized by the production of homocytotropic antibodies, which stimulate the release of pharmacologically active substances from the cells to which they bind. This type of immediate hypersensitivity can be detected by active and passive anaphylaxis. In the mouse, homocytotropic antibodies are of two types; one is a reaginic antibody (IgE) and the other is IgG_1. Inflammatory, cell-mediated immunity occurs in experimental trichinosis and it is passively transferable with the peritoneal exudate cells of infected mice. Acquired immunity to *Trichinella* is developed following a primary infection, so that the larvae of the challenge infection rapidly become eliminated in the immune host gut due to local inflammatory responses. Anti-inflammatory agents, such as cortisone, markedly depress the elimination of a challenge infection, as does whole-body-irradiation of hosts. Both anti-lymphocyte and anti-thymocyte sera reduce the resistance of immune mice to a challenge infection, implicating T-cells in acquired immunity.

Although the exact nature of the mechanisms of acquired resistance in laboratory animals to *Trichinella* remains to be elucidated, current opinion is that the parasite is expelled from the gut of the host by specific inflammatory mechanisms. It is possible that non-specific processes that alter the environment of the gut, rendering it unsuitable for the worms, are also involved. In trichinosis, protective immunity is directed essentially against invading larvae rather than established worms.

Ascaris suum in pigs and laboratory animals serves as a convenient model for human ascariasis (*A. lumbricoides*), the immunology of which is poorly understood. Immunity to *A. suum* possibly involves both humoral and cell-mediated elements, since passive transfer of immunity can sometimes be accomplished with anti-serum, but cannot be effected with lymphoid cells. Again, protective immunity acts against invading larvae, though self-cure has been recorded for adult ascarids in pigs. During an infection, IgM levels are the first to increase, followed by IgE, IgG and IgA. It is not known to which class of immunoglobulins the protective antibodies belong. In human infections with *Ascaris*, by contrast, the immunoglobulin titre does not always increase. Experimental ascariasis is typified by both inflammatory and cell-mediated responses directed, particularly, against larval worms in the tissues of the host. Granulomata are formed around parasite eggs, probably as a result of cell-mediated reactions. The third-stage larvae are the most active immunogenically and some characterization of parasite antigens has been accomplished by gel precipitation techniques.

Rather little is known about immunity to other nematodes that parasitize humans. *Ancylostoma caninum*, a dog hookworm, has been studied experimentally and is similar to *Nippostrongylus brasiliensis* in the rat, in the immune responses it invokes. Immunity to the human hookworms has not been examined in any detail. The immunology of human and experimental filarial infections has received some attention using *Litomosoides carinii* in cotton rats, *Dirofilaria immitis* in dogs and *Brugia pahangi* in cats as model systems. Infection in these systems produces increased IgE and IgG levels in the respective hosts. There is a significant degree of common antigenicity between filarial worms of different genera and other helminths.

Many nematodes, like some protozoans and platyhelminths, are capable of living in a host which has responded to the parasite by humoral and cellular reactions. The avoidance of the immune response is a characteristic feature of many parasitic infections, but little is known about the mechanisms involved. Additionally, there are unexpected interactions between nematodes and micro-organisms at the immunological level, e.g. mice infected with *Plasmodium berghei*, *Trypanosoma brucei*, or various babesias, lose their protective immune responses to *Trichuris muris* and *Trichinella spiralis*. No explanation of these events is, at present, forthcoming.

Immunization against parasitic diseases

The development of vaccines to combat any of the major parasitic diseases of man and domestic animals is of fundamental importance to the medical well-being and economy in much of the world, particularly the tropical and subtropical regions. Unfortunately, vaccine production has not yet met with any real success since so many parasites are capable of evading the immune responses of the host (antigenic variation in malaria and trypanosomiasis, and concomitant immunity in schisto-somiasis are examples of the problem). Nevertheless the search continues and it is a widely held view that a complete knowledge of the immune responses to each parasite is a prerequisite for the development of artificial immunization. As we have seen already in this chapter our depth of knowledge is far from complete for the immunology of any parasitic disease, but despite this, there have been encouraging results with immunization against some parasites.

A variety of different approaches has been adopted to produce immune hosts by artificial treatments and these include the use of attenuated parasites, homogenized parasite tissue, dead parasites, soluble parasite antigens, heterologous protection, controlled live infections and non-specific measures. Perhaps the most successful immunization yet achieved has come through the use of attenuated larvae to protect cattle against the nematode lungworm, *Dictyocaulus vivaparus*, and there is now a commercially available vaccine of irradiated larvae which confers a long-lasting immunity. Similar success has been achieved with *D. filaria* in sheep and *Ancylostoma caninum* in dogs. The use of larvae attenuated by irradiation has been found to offer protection in experimental infections of a number of nematode parasites, including *Brugia malayi*, *Dirofilaria immitis*, *Capillaria obsignata*, *Haemonchus contortus*, *Lito-mosoides carinii* and *Trichostrongylus colubriformis*, and in some schistosome species (*S. mattheei*). An effective live vaccine has been developed against *Babesia argentina* in Australia. The parasite is obtained from splenectomized calves and therefore has a lowered virulence and will not infect the tick vector. The use of killed parasites has not found much success and it has proved more efficacious to use soluble parasite antigens to induce immunity. Highly immunogenic substances have been isolated from a number of parasites, e.g. *Ascaris suum*, *Trichinella spiralis*, *Trichostrongylus colubriformis* and larvae of *Taenia saginata* and *Taenia ovis*. One of the more promising recent developments in the quest for suitable anti-parasite vaccines derives from tumour immunology. Following treatment with BCG (*Bacillus Calmette Guérin*) vaccine, non-specific immunity is developed for a wide range of parasites, e.g. experimental success has been achieved with rodent babesias, malarias, *Trypanosoma cruzi*, *Leishmania tropica* and *L. donovani*, *Echinococcus granulosus* and *E. multilocularis*. It now remains to be seen whether the potential of non-specific immunization with BCG vaccine and other such substances can be realized as a practical approach in the field.

Pathogenesis of parasitic infections

Immunopathology

It is well recognized that the pathology due to infection with many species of parasite is related to the immunological responses of the host to the parasite. Immunopathology is classified into four subclasses, that together constitute the allergic mechanisms often responsible for the production of disease symptoms.

Type I mechanisms: This class of allergic reactions includes immuno-pathology due to immediate hypersensitivity, in which parasite antigens react with host tissues that have been passively sensitized by antibodies, thus stimulating the release of pharmacologically active substances, which can, among other effects, increase blood supply and stimulate smooth muscle contraction in the host. The general term *anaphylaxis* is applied to the effects of these substances. Antibodies capable of sensitizing tissues for anaphylaxis are called *homocytotropic* antibodies and belong to the IgE class; IgG can also be involved. Included among the pharmacologically active substances that are released are histamine, serotonin, "slow reacting substance of anaphylaxis" and peptides called kinins (e.g. bradykinin).

Immediate hypersensitivity reactions have been identified in a number of parasitic infections including *Leishmania braziliiensis*, *Trichomonas foetus*, and *Trypanosoma cruzi*, as well as in many helminth infections, the latter involving both short-lived (IgG) and long-lived (IgE) homocyto-tropic antibodies. The prevalence of immediate hypersensitivity reactions in helminth infections provides a useful immuno-diagnostic tool for the identification of infected individuals, e.g. skin reactions, of the wheal and flare type, are often regarded as diagnostic for *Ascaris*, *Trichinella*, *Echinococcus*, *Schistosoma*, hookworms, filarial worms and many others.

In ascariasis, pathological conditions, including bronchitis, bronchial asthma and pulmonary eosinophilia, are associated with type I reactions. In trichinosis, the reactions include fever, skin rash, oedema and eosinophilia. Penetration of schistosome cercariae into sensitized mammalian skin is accompanied by urticaria ("nettle rash"), subcutaneous oedema, leucocytosis, eosinophilia and, occasionally, bronchial asthma. Infection with filarial worms is characterized by eosinophilia, while local swellings arise as a result of worm migration.

Type I reactions, occurring as they do in individuals with a previous experience of an infection (necessary to cause sensitization), are probably not involved directly in protective immunity. They may, however, affect the growth and development of parasites, and the ability of the parasite to migrate within the host.

Type II mechanisms This class of reactions involves the, usually, lytic effects of antibodies against antigens that are fixed to cell or basement membranes. The antigen may be an integral part of the membrane or it may simply be adsorbed to it. Complement and lymphocytes are both involved in type II reactions. It is thought that haemolytic anaemia in malaria results from this class of allergic reactions, as indeed may the anaemias associated with schistosomiasis, trypanosomiasis, leishmaniasis and babesiasis.

Type III mechanisms Immunopathology associated with this class of reactions is either inflammatory (arthus) or systemic (serum sickness), both of which result from the formation of antigen-antibody complexes. Complement is also involved in type III reactions. Arthus lesions are characterized by oedema, infiltration of neutrophils, necrosis of arteriole and venule walls, and thrombin-containing platelets and leucocytes. Serum sickness, due, for example, to the treatment of humans with animal serum, is rare nowadays. Pathological processes associated with this class of events include glomerulonephritis in malaria and schistsomiasis, and the nephrotic lesions in some malarias.

Type IV mechanisms These reactions include cell-mediated responses or delayed hypersensitivity. Delayed skin reactions have been demonstrated in human trypanosomiasis, leishmaniasis, toxoplasmosis, trichomoniasis, ascariasis, toxocariasis and schistosomiasis mansoni. In laboratory animals, many parasites induce delayed hypersensitivity reactions. The sores developed in cutaneous leishmaniasis (*Leishmania donovani*) and hepatosplenic disease (due to the formation of granulomata around tissue-bound eggs) in schistosomiasis are classical examples of type IV reactions.

General pathology

Apart from the immunopathology that we have discussed briefly above, many parasites cause damage to their hosts, resulting in the production of disease symptoms. Protozoan parasites multiply within the final host, whereas helminths do not. Therefore, any density-dependent effects of the former in the processes of pathogenesis will depend upon the rate of division or reproduction and in the latter on the rate of invasion, longevity and size of the parasites. These factors are of particular importance in pathogenesis. It must be made clear, however, that a great many parasites do not, under natural conditions, appear to cause their hosts any ill effects, i.e. an undetermined number of wild animals may be parasitized without ever showing outward or overt physiological signs of infection. It is rare, in the author's experience, to find unparasitized fishes, amphibians, birds or small mammals, yet these animals are living, until capture, apparently in harmony with their parasite fauna and it seems probable that it is only under certain circumstances that disease conditions arise from an infection. Such conditions would include the intensive monoculture of fishes, domestic fowl or cattle. Hence the view that a parasite is, by definition, a disease-causing organism is patently not true, and we must exercise care when defining the pathogenic nature of parasites, since only a relatively small number of species are disease-causing in natural environments. It has been an understandable tendency among parasitologists to focus attention on the pathogens and relegate the innocuous species to a lower order of significance.

Nevertheless, man, domestic and wild animals can be seriously affected by their parasite fauna. The ways in which pathology can arise are varied and are briefly outlined below.

Mechanical injury   There are a number of different ways in which parasites inflict mechanical injury to their hosts, e.g. during penetration and migration through the tissues, during feeding, through the action of the organs of attachment and cellular injury wrought by the presence of intracellular parasites within host cells. Many of the skin-invading digeneans produce localized damage at the site of entry, along with allergic reactions. Mechanical damage due to invasion and subsequent migration is often related to the number of invading larvae. The feeding habits of helminths, particularly nematodes, are responsible for considerable local damage to host epithelia, e.g. hookworms bite into the intestinal mucosa to obtain a blood and tissue meal and in doing so may leave open wounds with resultant bleeding and loss of tissue fluids. Mechanical damage attributable to parasite attachment organs is usually associated with the digeneans and cestodes. The unarmed tapeworm scolex, lacking hooks (e.g. Pseudophyllidea) may or may not inflict damage to the host mucosal tissue, whereas the presence of hooks on the scolex tends to be more injurious. In general, attachment of tapeworms to the intestinal mucosa is associated with inflammatory lesions and characteristic fibrosis. Digeneans, such as the liver flukes, produce pathological conditions by the mechanical disruption of host tissue, attributable to the body spines, attachment organs and the physical presence of the parasite in the narrow ducts of the liver. Anaemia, resulting from the ingestion of host red blood cells (e.g. *Schistosoma mansoni*, *Fasciola hepatica* and *Haemonchus contortus*) or from haemolysis (malaria) is not a common form of mechanical injury. Heavy infections of duct-dwelling parasites may cause occlusion of the ducts occupied, with its associated pathology, e.g. *Fasciola* in the bile duct, malaria in hepatic venules, and cestodes in the intestine. The severity of mechanical injury due to parasitic infection will usually be directly related to the number of parasites present, the size and behaviour of the parasite, or its multiplication rate if it is a protozoan.

Toxic effects   The most potent toxins known to man are produced by bacteria and include tetanus, diphtheria and botulism. By contrast, the majority of parasites do not release toxic substances into the blood and tissues of their hosts. Although toxins are thought to be produced by *Ascaris*, some digeneans (*Schistosoma*), some cestodes (*Moniezia*) and several protozoans (*Plasmodium* and *Sarcocystis*), there is no unequivocal evidence to support this notion. In malaria, for example, a toxin is postulated to inhibit mitochondrial respiration and oxidative phosphorylation in host liver cells, but this substance has not been isolated or identified.

Effects on host cell growth   Several parasites are responsible for inducing alterations in host cell and tissue growth patterns. Such effects are most commonly hyperplastic, i.e. cell growth without cell proliferation, or neoplastic, i.e. new cell growth, as in tumour production. Hyperplasia accompanies a number of infections, including *Eimeria, Schistosoma haematobium, Paragonimus westermani, Fasciola hepatica* and many adult tapeworm infections. Neoplasia, or tumour development, has been described for infections with *Taenia taeniaformis* in rodents, and *Schistosoma haematobium* in man.

Effects on the reproductive system of the host   There are a small number of well documented cases in which the host's reproductive system is altered, or its tissue destroyed, by the presence of a parasitic infection. The general term for this phenomenon is *parasitic castration*. The best known examples are *Sacculina parasitica* (a cirripede crustacean) that causes sex reversal in male crabs and sterility in females, and *Ligula intestinalis* (a pseudophyllidean cestode plerocercoid) which causes gonad regression and sterility in cyprinid fishes. In neither of these examples is the mechanism of castration understood. It seems probable that the parasites release compounds that mimic the activity of the host sex hormones, but extensive investigation has proved negative to date. The larvae of some digeneans are also known to castrate their molluscan intermediate hosts, e.g. *Himasthla secunda, Cercaria emaculens,* and *Cercaria lophocerca* in *Littorina littorea; Cercaria milfordensis* in *Mytilus edulis;* and *Leucochloridium paradoxum* in *Succinea putris.* The mechanisms involved here are also unknown.

Metabolic effects   Many parasites may bring about alterations in the various metabolic pathways of their hosts, but few cases have been documented. Among the best known are the pernicious anaemia and vitamin B_{12} deficiency characteristic of some human infections with the cestode *Diphyllobothrium latum,* and the growth-stimulating effects of larval *Spirometra mansonoides* (Pseudophyllidea) on laboratory rats and mice. In the latter case, the growth effect is due to the parasite releasing a substance (known as sparganum growth factor) that mimics the activities of mammalian growth hormone. This has questionable significance, however, since the larval parasite, the sparganum or plerocercoid, normally infects water snakes and is grown in rodents as a laboratory model only. Similar growth-stimulating effects are found in natural infections of molluscs with larval digeneans, and the phenomenon of molluscan parasitic gigantism is well known. Molluscs parasitized by some larval digeneans may show the reverse effect, i.e. stunting. Explanations for these interactions are not yet forthcoming.

SUMMARY

1. All animals are equipped with defence mechanisms to eliminate or control infecting organisms. Vertebrates employ non-specific phagocytic cells and specific antibodies; the latter do not feature in the defences of invertebrates. A brief description of the mammalian defence mechanisms is given.

2. Immunity to Protozoa is confused by the phenomenon of antigenic variation, by which process the parasite can evade the immune defences of the host by frequently changing its antigenic structure. Immunity to malarial parasites, trypanosomes and leishmanias has been studied extensively, but there is little promise of an effective vaccine against any of these important parasites at present. The relationship between *Trypanosoma lewisi* and ablastin offers an interesting model of acquired immunity. Premunition, whereby the initial infection is not eliminated by the immune host, is commonly associated with immunity to protozoan infections.

3. Immunity to helminths is characterized by an increase in the levels of circulating reaginic antibody (IgE). Schistosome infections are typified by concomitant immunity, which is thought to be a mechanism for evasion of the immune response. Immunity to tapeworm larvae may be protective, but in adult tapeworm infections it may control, rather than eliminate, the parasite. Some exceptions to this are described.

4. Self-cure (a dramatic expulsion of worms) occurs in several nematode infections, usually following an initial heavy infection. Inflammatory reactions are responsible for the expulsion of some nematodes.

5. Evasion of the immune response is a common feature of many parasitic infections. This fact presents a serious problem in the development of artificial vaccines against parasites. A few immunization procedures are commercially available and have achieved noteworthy success.

6. Four classes of immunopathogenesis are known in which the immune response of the host is responsible for producing a wide variety of disease symptoms. Other pathogenic mechanisms in parasitic infections include mechanical injury, toxic effects, growth effects (hyperplastic and neoplastic), effects on the host's reproductive system and various metabolic changes.

FURTHER READING

General

Baer, J. G. (1971) *Animal Parasites*, World University Library, London.
Baker, J. R. (1969) *Parasitic Protozoa*, Hutchinson University Library.
Chandler, A. C. and Read, C. P. (1960) *Introduction to Parasitology*, Wiley.
Cheng, T. C. (1973) *General Parasitology*, Academic Press.
Crofton, H. D. (1966) *Nematodes*, Hutchinson University Library.
Kennedy, C. R. (1975) *Ecological Animal Parasitology*, Blackwell.
Read, C. P. (1970) *Parasitism and Symbiology*, Ronald Press.
Read, C. P. (1972) *Animal Parasitism*, Prentice Hall.
Schmidt, G. D. and Roberts, L. S. (1977) *Foundations of Parasitology*, Mosby.

Chapter 2

Bennet-Clark, H. C. (1976) "Mechanics of nematode feeding", in *The Organisation of Nematodes*, editor N. A. Croll, Academic Press, 313–342.
Clegg, J. A. and Smyth, J. D. (1968) "Growth, development and culture methods: parasitic platyhelminths", in *Chemical Zoology II. Porifera, Coelenterata and Platyhelminthes*, editors M. Florkin and B. T. Scheer, Academic Press, 395–466.
Conner, R. L. (1967) "Transport phenomena in Protozoa", in *Chemical Zoology I. Protozoa*, editor G. W. Kidder, Academic Press, 309–350.
Dewey, V. C. (1967) "Lipid composition, nutrition and metabolism", in *Chemical Zoology I. Protozoa*, editor G. W. Kidder, Academic Press, 93–161.
Erasmus, D. A. (1977) "The host-parasite interface of Trematoda." *Advances in Parasitology*, **15**, 201–243.
Hockley, D. A. (1973) "Ultrastructure of the tegument of *Schistosoma*." *Advances in Parasitology*, **11**, 233–305.
Jennings, J. B. (1968) "Nutrition and digestion", in *Chemical Zoology II. Porifera, Coelenterata and Platyhelminthes*, editors M. Florkin and B. T. Scheer, Academic Press, 305–327.
Kidder, G. W. (1967) "Nitrogen: distribution, nutrition and metabolism", in *Chemical Zoology I. Protozoa*, editor G. W. Kidder, Academic Press, 93–161.
Lee, D. L. (1972) "The structure of the helminth cuticle." *Advances in Parasitology*, **10**, 347–379.
Mettrick, D. F. and Podesta, R. B. (1974) "Ecological and physiological aspects of helminth-host interactions in the mammalian gastrointestinal canal." *Advances in Parasitology*, **12**, 183–279.
Pappas, P. W. and Read, C. P. (1975) "Membrane transport in helminth parasites: a review." *Experimental Parasitology*, **37**, 469–530.
Read, C. P. (1971) "The microcosm of intestinal helminths", in *Ecology and Physiology of Parasites*, editor A. M. Fallis, Adam Hilger Ltd., 188–200.
Read, C. P., Rothman, A. H. and Simmons, J. E. (1963) "Studies on membrane transport, with special reference to parasite-host integration." *Annals of the New York Academy of Science*, **113**, 154–205.

Smyth, J. D. (1977) *Introduction to Animal Parasitology*, 2nd edition, Hodder and Stoughton.

Taylor, A. E. R. and Baker, J. R. (1968) *The Cultivation of Parasites in vitro*. Blackwell.

Von Brand, T. (1973) *Biochemistry of Parasites*, 2nd edition, Academic Press.

Chapter 3

Atkinson, H. J. (1976) "The respiratory physiology of nematodes", in *The Organization of Nematodes*, editor N. A. Croll, Academic Press, 243–272.

Barrett, J. (1976) "Bioenergetics in helminths", in *Biochemistry of Parasites and Host-Parasite Relations*, editor H. Van den Bossche, North Holland, 67–80.

Bryant, C. (1970) "Electron transport in parasitic helminths and Protozoa." *Advances in Parasitology*, **8**, 139–172.

Bryant, C. (1975) "Carbon dioxide utilisation in parasitic helminths." *Advances in Parasitology*, **13**, 35–69.

Coles, G. C. (1973) "The metabolism of *Schistosoma*: a review." *International Journal of Biochemistry*, **4**, 319–337.

Coles, G. C. (1975) "Fluke biochemistry—*Fasciola* and *Schistosoma*." *Helminthological Abstracts (A)*, **44**, 147–162.

Gutteridge, W. E. and Coombs, G. H. (1977) *Biochemistry of Parasitic Protozoa*, Macmillan.

Hill, G. C. (1970) "Characterization of the electron transport systems present during the life cycle of African trypanosomes", in *Biochemistry of Parasites and Host-Parasite Relations*, editor H. Van den Bossche, North Holland, 31–50.

Hill, G. C. and Anderson, W. A. (1970) "Electron transport systems and mitochondrial DNA in Trypanosomatidae: a review." *Experimental Parasitology*, **28**, 356–380.

Lee & Atkinson (1976) *Physiology of Nematodes*, 2nd edition, Macmillan Press.

Read, C. P. & Simmons, J. E. (1963) "Biochemistry and physiology of tapeworms." *Physiological Reviews*, **43**, 263–305.

Saz, H. J. (1969) "Carbohydrate and energy metabolism of nematodes and Acanthocephala." *Chemical Zoology*, **3**, 329–360.

Smith, M. H. (1969) "Do intestinal parasite require oxygen?" *Nature*, **223**, 1129–1132.

Von Brand, T. (1973) *Biochemistry of Parasites*, 2nd edition, Academic Press.

Chapter 4

Borst, P. and Fairlamb, A. H. (1976) "DNA of parasites, with special reference to kinetoplast DNA," in *Biochemistry of Parasites and Host-Parasite Relations*, editor H. Van den Bossche. North Holland Publishing Co., 169–192.

Dewey, V. C. (1967) "Lipid composition, nutrition and metabolism", in *Chemical Zoology*, *I: Protozoa*, editor G. W. Kidder. Academic Press, 162–274.

Gutteridge, W. E. and Coombs, G. H. (1977) *Biochemistry of Parasitic Protozoa*, Macmillan Press.

Kidder, G. W. (1967) " Nitrogen: distribution, nutrition and metabolism", in *Chemical Zoology*, *I: Protozoa*, editor G. W. Kidder. Academic Press, 93–161.

Lee, D. L. and Smith, M. H. (1965) "Haemoglobins of parasitic animals." *Experimental Parasitology*, **16**, 394–424.

Mandel, M. (1967) "Nucleic acids of Protozoa", in *Chemical Zoology*, *I: Protozoa*, editor G. W. Kidder. Academic Press, 541–573.

Meyer, F. and Meyer, H. (1972) "Loss of fatty acid biosynthesis in flatworms", in *Comparative Biochemistry of Parasites*, editor H. Van den Bossche, Academic Press, 383–393.

Newton, B. A. (1973) "Trypanocides as biochemical probes", in *Chemotherapeutic Agents in the Study of Parasites*, Symposia of the British Society for Parasitology, **11**, editors A. E. R. Taylor and R. Muller, 29–51.

Newton, B. A. and Burnett, J. K. (1972) "DNA of Kinetoplastidae: a comparative study", in *Comparative Biochemistry of Parasites*, editor H. Van den Bossche, Academic Press, 185–198.

Newton, B. A., Cross, G. A. M. and Baker, J. R. (1973) "Differentiation in the Trypanosomatidae", in *Microbial Differentiation*, Symposia of the Society for General Microbiology, **23**, 339–373.

Read, C. P. (1968) "Intermediary metabolism of flatworms. in *Chemical Zoology, II: Porifera, Coelenterata and Platyhelminths*", editors M. Florkin and B. Scheer, Academic Press, 328–357.

Von Brand, T. (1973) *Biochemistry of Parasites*, 2nd edition, Academic Press.

Chapter 5

Campbell, J. C. (1963) "Urea formation and urea cycle enzymes in the cestode, *Hymenolepis diminuta*." *Comparative Biochemistry and Physiology*, **8**, 13–27.

Howells, R. E. (1969) "Observations on the nephridial system of the cestode *Moniezia expansa*." *Parasitology*, **59**, 449–459.

Kidder, G. W. (1967) "Nitrogen: distribution, nutrition and metabolism", in *Chemical Zoology* 1. *Protozoa*, editor G. W. Kidder, Academic Press.

Lee, D. L. and Atkinson, H. J. (1976) *Physiology of Nematodes*, 2nd edition, Macmillan Press.

Read, C. P. and Simmons, J. E. (1963) "Biochemistry and physiology of tapeworms." *Physiological Reviews*, **43**, 263–305.

Siddiqi, A. H. and Lutz, P. L. (1966) "Osmotic and ionic regulation in *Fasciola gigantica* (Trematoda: Digenea)." *Experimental Parasitology*, **19**, 348–357.

Von Brand, T. (1973) *Biochemistry of Parasites*, 2nd edition, Academic Press.

Webster, L. A. (1971) "The flow of fluid in the protonephridial canals of *Hymenolepis diminuta*." *Comparative Biochemistry and Physiology*, **39A**, 785–793.

Webster, L. A. and Wilson, R. A. (1970) "The chemical composition of protonephridial canal fluid from the cestode *Hymenolepis diminuta*." *Comparative Biochemistry and Physiology*, **35**, 201–209.

Wilson, R. A. (1967) "The protonephridial system in the miracidium of the liver fluke *Fasciola hepatica*." *Comparative Biochemistry and Physiology*, **20**, 337–342.

Wilson, R. A. (1969) "The fine structure of the protonephridial system in the miracidium of *Fasciola hepatica*." *Parasitology*, **59**, 461–467.

Wilson, R. A. and Webster, L. A. (1974) "Protonephridia." *Biological Reviews*, **49**, 127–160.

Wright, D. J. and Newell, D. P. (1976) "Nitrogen excretion, osmotic and ionic regulation in nematodes", in *The Organisation of Nematodes*, editor N. A. Croll, Academic Press.

Chapter 6

Bychowsky, B. E. (1957) *Monogenetic Trematodes*, American Institute of Biological Sciences.

Crompton, D. W. T. (1970) *An Ecological Approach to Acanthocephalan Physiology*, Cambridge University Press.

Erasmus, D. A. (1972) *The Biology of Trematodes*, Edward Arnold.

Grell, K. G. (1967) "Sexual reproduction in Protozoa", in *Research in Protozoology*, **2**, editor T. T. Chen, 148–213.

Hanson, E. D. (1967) "Protozoan development", in *Chemical Zoology*, **1**, *Protozoa*, editor G. W. Kidder, 395–539.

Lee, D. L. and Atkinson, H. J. (1976) *Physiology of Nematodes*, 2nd edition, Macmillan Press.

Llewellyn, J. (1972) "Behaviour of monogeneans", in *Behavioural Aspects of Parasite Transmission*, Linnean Society (supplement 51), editors E. U. Canning and C. A. Wright, 19–30.

Schmidt, G. D. and Roberts, L. S. (1977) *Foundations of Parasitology*, C. V. Mosby.

Smyth, J. D. (1966) *The Physiology of Trematodes*, Oliver and Boyd.

Smyth, J. D. (1969) *The Physiology of Cestodes*, Oliver and Boyd.

Smyth, J. D. (1976) *Introduction to Animal Parasitology*, 2nd edition, Hodder and Stoughton.

Chapter 7

Cable, R. M. (1972) "Behaviour of digenetic trematodes", in *Behavioural Aspects of Parasite Transmission*, Linnean Society, editors E. U. Canning & C. A. Wright, Academic Press, 1–18.

Cheng, T. C. (1967) "Marine molluscs as hosts for symbioses", in *Advances in Marine Biology*, **5**, editor F. S. Russell, 16–134.

Hawking, F. (1975) "Circadian and other rhythms of parasites." *Advances in Parasitology*, **13**, 123–182.

Kearn, G. C. (1970) "The physiology and behaviour of the monogenean skin parasite *Entobdella soleae* in relation to its host (*Solea solea*)", in *Ecology and Physiology of Parasites*, editor A. M. Fallis, Adam Hilger Ltd., 161–187.

Lee, D. L. and Atkinson, H. J. (1976) *Physiology of Nematodes*, 2nd edition, Macmillan Press.

Llewellyn, J. (1972) "Behaviour of monogeneans", in *Behavioural Aspects of Parasite Transmission*, Linnean Society, editors E. U. Canning and C. A. Wright, Academic Press, 19–30.

Matthews, B. E. (1977) "The passage of larval helminths through tissue barriers", in *Parasite Invasion*, Symposia of the British Society for Parasitology, **15**, editors A. E. R. Taylor and R. Muller, 93–120.

MacInnis, A. J. (1965) "Responses of *Schistosoma mansoni* miracidia to chemical attractants." *Journal of Parasitology*, **51**, 731–746.

Read, C. P. (1970) *Parasitism and Symbiology*, Ronald Press.

Smyth, J. D. (1966) *The Physiology of Trematodes*, Oliver and Boyd.

Stirewalt, M. A. (1966) "Skin penetration mechanisms in helminths", in *Biology of Parasites*, editor E. J. L. Soulsby, Academic Press, 41–60.

Ulmer, M. J. (1970) "Site finding behaviour in helminths in intermediate and definitive hosts", in *Ecology and Physiology of Parasites*, editor A. M. Fallis, Adam Hilger Ltd., 121–160.

Chapter 8

Bannister, L. H. (1977) "The invasion of red cells by *Plasmodium*", in *Parasite Invasion*, Symposia of the British Society for Parasitology, **15**, editors A. E. R. Taylor and R. Muller, 27–55.

Barrett, J. (1977) "Energy metabolism and infection in helminths", in *Parasite Invasion*, Symposia of the British Society for Parasitology, **15**, editors A. E. R. Taylor and R. Muller, 121–144.

Croll, N. & Matthews, B. E. (1977) *Biology of Nematodes*, Blackie and Son.

Crompton, D. W. T. (1973) "The sites occupied by some parasitic helminths in the alimentary tract of vertebrates." *Biological Reviews*, **48**, 27–83.

Erasmus, D. A. (1972) *The Biology of Trematodes*, Edward Arnold.

Holmes, J. C. (1973) "Site selection by parasitic helminths: interspecific interactions, site segregation, and their importance to the development of helminth communities." *Canadian Journal of Zoology*, **51**, 333–347.

Lackie, A. M. (1975) "The activation of infective stages of endoparasites of vertebrates." *Biological Reviews*, **50**, 285–323.

Lee, D. L. and Atkinson, H. J. (1976) *The Physiology of Nematodes*, 2nd edition, Macmillan Press.

Llewellyn, J. (1976) "Behaviour of monogeneans", in *Behavioural Aspects of Parasite Transmission*, Linnean Society, editors E. U. Canning and C. A. Wright, 19–30.

Otto, G. F. (1966) "Development of parasitic stages of nematodes", in *Biology of Parasites*, editor E. J. L. Soulsby, Academic Press, 85–99.

Read, C. P. and Kilejian, A. Z. (1969) "Circadian migratory behaviour of a cestode symbiote in the rat host." *Journal of Parasitology*, **55**, 574–578.

Roberts, L. S. (1961) "The influence of population density on patterns and physiology of growth in *Hymenolepis diminuta* (Cestoda: Cyclophyllidea) in the definitive host." *Experimental Parasitology*, **11**, 332–371.

Rogers, W. P. (1966) "Exsheathment and hatching mechanisms in helminths", in *Biology of Parasites*, editor E. J. L. Soulsby, Academic Press, 33–40.

Schad, G. A. (1963) "Niche diversification in a parasite species flock. "*Nature*, **198**, 404–406.

Smyth, J. D. (1969) *The Physiology of Cestodes*, Oliver and Boyd.

Smyth, J. D. (1969) "Parasites as biological models." *Parasitology*, **59**, 73–92.

Smyth, J. D. (1976) *Introduction to Animal Parasitology*, 2nd edition, Hodder and Stoughton.

Smyth, J. D. and Haslewood, G. A. D. (1963) "The biochemistry of bile as a factor determining host specificity in intestinal parasites, with particular reference to *Echinococcus granulosus*." *Annals of the New York Academy of Science*, **113**, 234–260.

Ulmer, M. J. (1971) "Site finding behaviour in helminths in intermediate and definitive hosts", in *Ecology and Physiology of Parasites*, editor A. M. Fallis, Adam Hilger Ltd., 123–160.

Chapter 9

Brooker, B. E. (1972) "The sense organs of trematode miracidia", in *Behavioural Aspects of Parasite Transmission*, Linnean Society, editors E. U. Canning and C. A. Wright, Academic Press, 171–180.

Bychowsky, B. E. (1957) *Monogenetic Trematodes*, American Institute of Biological Science.

Croll, N. A. (1975) "Behavioural analyses of nematode movement." *Advances in Parasitology*, **13**, 71–122.

Croll, N. A. (1976) "Behavioural coordination of nematodes", in *The Organization of Nematodes*, editor N. A. Croll, Academic Press, 343–364.

Erasmus, D. A. (1972) *The Biology of Trematodes*, Edward Arnold.

Hockley, D. J. (1973) "Ultrastructure of the tegument of *Schistosoma*." *Advances in Parasitology*, **11**, 233–305.

Hymen, L. H. (1951) *The Invertebrates, Acanthocephala, Aschelminthes and Entoprocta*. McGraw-Hill.

Isseroff, H. and Cable, R. M. (1968) "The fine structure of photoreceptors in larval trematodes." *Zeitschrift fur Zellforschung*, **86**, 511–534.

Jarman, M. (1976) "Neuromuscular physiology of nematodes", in *The Organization of Nematodes*, editor N. A. Croll, Academic Press, 293–312.

Lee, D. L. and Atkinson, H. J. (1976) *Physiology of Nematodes*, 2nd edition, Macmillan Press.

Lyons, K. M. (1972) "Sense organs of monogeneans", in *Behavioural Aspects of Parasite Transmission*, Linnean Society, editors E. U. Canning and C. A. Wright, Academic Press, 181–200.

Lyons, K. M. (1973) "The epidermis and sense organs of the Monogenea and some related groups." *Advances in Parasitology*, **11**, 193–232.

McLaren, D. J. (1976) "Nematode sense organs." *Advances in Parasitology*, **14**, 195–265.

Nicholas, W. L. (1973) "The biology of the Acanthocephala." *Advances in Parasitology*, **11**, 671–706.

Chapter 10

Bawden, M. P. (1975) "Whence comes *Trypanosoma lewisi* antigen which induces ablastic antibody: studies in the occult?" *Experimental Parasitology*, **38**, 350–356.

Capron, A., Dessaint, J. P., Camus, D. and Capron, M. (1976) "Some immune mechanisms in host parasite relationship", in *Biochemistry of Parasites and Host-Parasite Relationships*, editor H. Van den Bossche, North Holland, 263–282.

Cohen, S. and Sadun, E. H. (1976) *Immunology of Parasitic Infections*, Blackwell.

Cox, F. E. G. (1968) "Immunity to tissue Protozoa", in *Immunity to Parasites*, Syposium of the British Society for Parasitology, **6**, editor A. E. R. Taylor, 5–23.

Cox, F. E. G. (1978) "Specific and non specific immunisation against parasitic infections." *Nature*, **273**, 623–626.

D'Alessandro, P. A. (1975) "Ablastin: the phenomenon." *Experimental Parasitology*, **38**, 303–308.

Dusanic, D. G. (1975) "Immunosuppression and ablastin." *Experimental Parasitology*, **38**, 322–337.

Gray, A. R. and Luckins, A. G. (1976) "Antigenic variation in salivarian trypanosomes", in *Biology of the Kinetoplastida*, editors W. H. R. Lumsden and D. A. Evans, Academic Press, 493–542.

Larsh, J. E. and Weatherly, N. F. (1975) "Cell mediated immunity against certain parasitic worms." *Advances in Parasitology*, **13**, 183–224.

Maegraith, B. and Fletcher, A. (1972) "The pathogenesis of mammalian malaria." *Advances in Parasitology*, **10**, 49–75.

Ormerod, W. E. (1975) "Ablastin in *Trypanosoma lewisi* and related phenomena in other species of trypanosomes." *Experimental Parasitology*, **38**, 338–341.

Read, C. P. (1970) *Parasitism and Symbiology*, Ronald Press.

Rees, G. (1967) "Pathogenesis of adult cestodes." *Helminthological Abstracts*, **36**, 1–23.

Roitt, I. (1977) *Essential Immunology*, 3rd edition, Blackwell.

Sinclair, I. J. (1970) "The relationship between circulating antibodies and immunity to helminthic infections." *Advances in Parasitology*, **8**, 97–138.

Sinclair, K. B. (1967) "Pathogenesis of *Fasciola* and other liver flukes." *Helminthological Abstracts*, **36**, 115–134.

Smithers, S. R. and Terry, R. J. (1976) "The immunology of schistosomiasis." *Advances in Parasitology*, **14**, 399–422.

Smyth, J. D. (1977) *Introduction to Animal Parasitology*, 2nd edition, Hodder and Stoughton.

Smyth, J. D. and Heath, D. D. (1970) "Pathogenesis of larval cestodes in mammals." *Helminthological Abstracts*, **39**, 1–23.

Targett, G. A. T. (1968) "Acquired immunity to blood parasitic Protozoa", in *Immunity to Parasites*, Symposium of the British Society for Parasitology, **6**, editor A. E. R. Taylor, 25–41.

Vickerman, K. (1978) "Antigenic variation in trypanosomes." *Nature*, **273**, 613–617.

Warren, K. S. (1973) "The pathology of schistosome infections." *Helminthological Abstracts*, **42**, 592–633.

Warren, K. S. (1978) "The pathology, pathobiology and pathogenesis of schistosomiasis." *Nature*, **273**, 609–612.

Wright, C. A. (1966) "The pathogenesis of helminths in the Mollusca." *Helminthological Abstracts*, **35**, 207–224.

Zuckerman, A. (1975) "Current status of the immunology of blood and tissue Protozoa. I *Leishmania*." *Experimental Parasitology*, **38**, 370–400.

Zuckerman, A. (1977) "Current status of the immunology of blood and tissue Protozoa. II *Plasmodium*." *Experimental Parasitology*, **42**, 374–446.

see also: *Immunology of Parasitic Infections*, Workshop Report, American Journal of Tropical Medicine and Hygiene, **26**, (1977).

GLOSSARY

Agglutination (in immunology)   cross-linking of cells by antibodies directed against surface antigens causing clumping of invading cells.

Allosteric mechanisms (in enzymology and membrane transport)   inhibition or stimulation of enzyme activity or a transport process by the noncompetitive binding of molecules at sites other than the active site of the enzyme or the "carrier" of a transport system.

Anastamosis   a branch; usually refers to a branched canal system, e.g. the gut.

ATP   adenosine triphosphate; an energy-rich molecule which is hydrolyzed to the di- and monophosphate to provide energy.

Babesiasis   the group of diseases caused by sporozoans of the genus *Babesia* and transmitted to vertebrates by ticks.

Bifid   divided into two parts.

Bipolar body   rod-shaped cytoplasmic structures found in *Crithidia oncopelti* and now regarded by most workers as endosymbiotic bacteria.

Cercaria   the second larval stage of digeneans, which is released from snails and locates the next host in the life-cycle, movement by swimming or crawling.

Chemotherapy   the treatment of any disease with synthetic drugs.

Coccidiosis   A rather arbitrary name for the diseases caused by certain members of the Coccidia; usually refers to infection with the genera *Eimeria, Isospora* and *Lankasterella*.

Complement (in immunology)   a series of nine serum proteins that act sequentially after activation by antibody; the net result is the lysis of invading cells. Complement fixation provides a useful immunodiagnostic tool for detecting many parasitic infections.

Coracidium   a ciliated, free-swimming larval tapeworm.

Cysticercoid   an encysted larval tapeworm, common in the Cyclophyllidea.

Cysticercosis   infection of mammals with the cysticercus larva of tapeworms, usually belonging to the genus *Taenia*.

Cytostome   a distinct, mouth-like structure on the surface of some protozoans through which large nutrient molecules may be obtained.

Cytotoxicity (in immunology)   lysis of foreign cells following complement fixation by antibodies directed against cell surfaces.

Echinostomes   an order of digeneans in which the cercaria and adult frequently possess a circum-oral collar armed with spines.

Embryophore   a thickened membrane surrounding a larval tapeworm within the egg; closely associated with the inner envelope.

Endoplasmic reticulum   membranous structures in the cytosol of many cells, highly developed in protein-secreting cells when associated with ribosomes forming rough or granular endoplasmic reticulum.

Exflagellation   in malaria parasites, prior to fertilisation, the male gametocyte develops eight flagella which emerge through the surface of the gametocyte.

Fascioliasis   liver fluke disease of cattle, sheep and man caused by the digeneans, *Fasciola* and related genera.

Filariasis   the group of diseases of man and animals caused by the filarial nematodes, including elephantiasis and onchocerciasis (river blindness) among others.

Gamogony   the sexual process of gamete production, especially in the Protozoa.

211

Gastrodermis the epithelial lining of the alimentary canal of invertebrates.

Golgi apparatus a complex of vesicles and membranes found in many cells (named after its discoverer, Camillo Golgi); involved in the production of secretory bodies.

Gynaecophoric canal a groove running the length of the male schistosome, in which the female worm resides.

Haemoflagellates blood-dwelling flagellate Protozoa, including the trypanosomes and related forms.

Hapten a small molecule, which by itself is non-immunogenic, but which can combine with antibody; may induce immune responses when combined with carrier molecules.

Helminth a general term used to include the Platyhelminths, the Nematoda and the Acanthocephala.

Helminthiasis an infection with a helminth parasite, not necessarily resulting in disease symptoms.

Homocytotropic antibodies reagin (IgE); antibodies that, on contact with antigen, cause the release of vasoactive substances and histamine from mast cells.

In vitro culture the artificial growth and maintenance of parasites outside the body of the natural host.

In vivo culture maintenance of parasites within the body of a host animal or plant.

Kinetoplast a distinct region of the mitochondrion of trypanosomes, containing unique DNA.

Leishmaniasis the group of diseases caused by protozoans of the genus *Leishmania*, including kala-azar and oriental sore.

Logical anthelminthics synthetic anti-parasitic drugs developed from a knowledge of the biochemical pathways of parasites; an ideal rarely achieved.

Lymphokines mediating factors in delayed hypersensitivity reactions which are released from T-lymphocytes after contact with antigens.

Lysis the process of splitting or breaking down; as in haemolysis—the rupture of red blood cells; proteolysis—the enzymic degradation of proteins.

Lysosome a granule with lytic capabilities, common in the cytosol of phagocytic cells; arises from the Golgi apparatus.

Malaria the disease of birds and mammals caused by the blood-dwelling protozoans of the genus *Plasmodium*; the parasites invade red blood cells and cause a range of disease symptoms. Human malaria is a tropical disease of major importance.

Metacercaria the larval stage in the life-cycle of many digeneans that immediately precedes the adult; may or may not be encysted and parasitic.

Micelle a complex of lipid molecules in solution, in which the hydrophobic ends of the molecules are directed away from the aqueous phase.

Microtriches the microvilli-like projections on the surface of tapeworms; characterized by their electron-dense tips of unknown function.

Miracidium the first larval stage of the Digenea; a ciliated, swimming larva that emerges from the egg and locates and penetrates the snail host.

Naive (in immunology) an animal with no previous experience of infection with a particular parasite.

Neoteny the development of precocious sexual maturity in a larval form.

Oncomiracidium the ciliated, free-swimming larva of the Monogenea.

Oncosphere a 6-hooked, tapeworm larval stage ($\equiv$ coracidium).

Opsonising antibodies (*opsonins*) antibodies that bind to foreign cell surfaces and enhance phagocytosis by macrophages.

Peyer's Patches nodules of lymphoid tissue in the wall of the mammalian intestine, concerned with protection against intestinal parasites.

Photoreceptors light-sensitive sense organs (eye-spots, eyes, etc.).

Pinocytosis the passage of macromolecules across surface membranes involving vesicle formation.

Plerocercoid a progenetic (advanced) larval pseudophyllidean tapeworm; sometimes termed a sparganum.

Polymorph polymorphonuclear leucocytes.

Precipitation the formation of a visible complex when soluble antigen and soluble antibody are mixed; such complexes play an important role in immuno-diagnosis.

Premunition a state of protective immunity that depends upon the presence of the invading parasite; complete clearance of all parasites will render the host susceptible to reinfection.

Procercoid a tapeworm larva developed from a coracidium, usually with a posterior cercomer or tail-like structure; a parasitic larva.

Proprioreceptor sensory receptor that provides information on internal conditions, e.g. a stretch receptor.

Protandry in hermaphrodites, the male reproductive system develops before the female system.

Recrudescence in malaria, short-term recurrence of the disease due to the survival of intraerythrocytic forms of the parasite; *relapse* is the long-term recurrence due to the survival of exoerythrocytic forms.

Redia an intramolluscan larval stage of the Digenea.

Redox the oxidation-reduction state.

Rheoreceptor sense organ that responds to directional water currents.

Ribosome intracellular unit concerned with protein synthesis.

Salivaria trypanosomes usually transmitted by tsetse flies, in which the parasite inhabits the fore-gut and salivary glands; transmission is via the bite of the vector, e.g. *Trypanosoma*.

Sanguinivorous blood feeding.

Schistosomiasis the disease caused by blood flukes of the genus *Schistosoma* and related genera; the three human forms (*S. mansoni*, *S. haematobium* and *S. japonicum*) are a major tropical problem.

Schizogony asexual reproduction in the Protozoa to form schizonts or merozoites.

Scolex the anterior attachment organ of adult tapeworms.

Sporocyst 1. an intramolluscan larval stage of the Digenea developed from the miracidium; occurs first as mother form, from which develop daughter forms.
2. a cystic form of some protozoans that contains sporozoites.

Sporogony asexual reproduction in many protozoans to form sporozoites.

Stercoraria trypanosomes that develop in the hind-gut of the insect vector; transmission occurs with the faeces of the host during its feeding activities, e.g. *Schizotrypanum*.

Sterile immunity immunity that results in the complete elimination of the infection.

Strigeid referring to digeneans of the superfamily Strigeoidea.

Strobila the segmented or proglottised body of an adult tapeworm.

Taeniasis the disease caused by cestodes of the genus *Taenia*, e.g. *T. solium*, the pork tapeworm of man.

Tangoreceptor touch receptor.

Taxis movement towards or away from a particular stimulus, e.g. phototaxis, geotaxis, rheotaxis.

Theilerasis diseases of cattle, sheep and goats caused by the tick-borne protozoans of the genus *Theileria*.

Trypanosomiasis disease caused by haemoflagellate protozoans of the genus *Trypanosoma* and related genera, e.g. human sleeping sickness, Chagas' disease.

APPENDIX

AN OUTLINE CLASSIFICATION OF PARASITES

Phylum Protozoa
  Subphylum Sarcomastigophora
    Superclass Mastigophora
      Class Zoomastigophora
          Order Rhizomastigida (*Histomonas*)
          Order Kinetoplastida
            Suborder Bodonina (*Cryptobia*)
            Suborder Trypanosomatina (*Trypanosoma, Leishmania*)
          Order Retortomonadida
          Order Diplomonadida (*Giardia*)
          Order Oxymonadida
          Order Trichomonadida (*Trichomonas*)
          Order Hypermastigida (*Trichomympha*)
    Superclass Opalinata (*Opalina*)
    Superclass Sarcodina (*Entamoeba, Hartmanella*)
  Subphylum Apicomplexa
      Class Sporozoa
        Subclass Gregarinasina (*Monocystis*)
        Subclass Coccidiasina
          Order Protococcidiorida
          Order Eucoccidiorida
            Suborder Adeleorina (*Hepatozoon*)
            Suborder Eimeriorina
              Family Selenococcidiidae
              Family Aggregatidae
              Family Caryotrophidae
              Family Lankasterellidae (*Lankasterella*)
              Family Eimeriidae (*Eimeria, Isospora*)
              Family Cryptosporidiidae
              Family Pfeifferinellidae
              Family Sarcocystidae
                Subfamily Toxoplasmatinae (*Toxoplasma*)
                Subfamily Besnoitiinae
                Subfamily Sarcocystinae (*Sarcocystis*)
            Suborder Haemosporina (*Plasmodium, Haemoproteus*)
      Class Piroplasmida (*Babesia, Theileria*)
  Subphylum Cnidospora
      Class Myxosporidea
      Class Microsporidea

215

Phylum Protozoa—*continued*
  Subphylum Ciliophora
    Class Cileata
      Subclass Holotricha (*Balantidium, Ichthyophthirius*)
      Subclass Peritricha (*Trichodina*)
      Subclass Suctoria
      Subclass Spirotricha (*Nyctotherus, Entodinium*)

Phylum Platyhelminthes
    Class Monogenea
        Order Monopisthocotylea (*Gyrodactylus*)
        Order Polyopisthocotylea (*Polystoma, Entobdella, Diclidophora*)
    Class Digenea
      Subclass Gasterostomata (*Bucephalus*)
      Subclass Prosostomata
        Order Strigeata
          Superfamily Strigeoidea (*Diplostomum*)
          Superfamily Schistosomatoidea (*Schistosoma*)
          Superfamily Azygoidea
          Superfamily Transversotrematoidea
        Order Echinostomata
          Superfamily Echinostomatoidea (*Fasciola*)
          Superfamily Paramphistomoidea
          Superfamily Notocotyloidea (*Notocotylus*)
        Order Plagiorchiata
          Superfamily Plagiorchioidea (*Haematoloechus, Haplometra*)
          Superfamily Allocreadioidea (*Allocreadium*)
        Order Opisthorchiata
    Class Cestoidea
      Subclass Cestodaria
        Order Amphilinidea
        Order Gyrocotylidea
      Subclass Eucestoda
        Order Tetraphyllidea (*Phyllobothrium*)
        Order Proteocephalidea (*Proteocephalus*)
        Order Caryophyllidea (*Caryophyllaeus*)
        Order Trypanorhyncha (*Lacistorhynchus*)
        Order Pseudophyllidea (*Ligula, Diphyllobothrium*)
        Order Cyclophyllidea (*Taenia, Hymenolepis, Echinococcus*)

Phylum Nematoda
    Class Aphasmidea
        Order Trichinellida (*Trichuris, Trichinella*)
        Order Dictyophymatida
    Class Phasmidea
        Order Rhabdidata (*Rhabdias*)
        Order Strongylata (*Ancylostoma*)
        Order Ascaridata (*Ascaris*)
        Order Oxyurata (*Heterakis*)
        Order Spirurata (*Tetrameres*)
        Order Camallanata (*Camallanus, Dracunculus*)
        Order Filariata (*Wuchereria, Loa, Brugia*)

Phylum Acanthocephala
    Class Archiacanthocephala (*Moniliformis*)
    Class Palaeacanthocephala (*Polymorphus*)
    Class Eoacanthocephala (*Neoechinorhynchus*)

It is generally accepted that the Platyhelminthes, the Nematoda and the Acanthocephala are loosely grouped together under the name of "helminths".

Other invertebrate phyla that contain parasitic members include: Arthropoda (Crustacea, Insecta, Arachnida); Porifera; Coelenterata; Ctenophora; Mesozoa; Nemertea; Rotifera; Nematomorpha; Annelida; Pentastomida and Mollusca.

INDEX

ablastin   184
Acanthocephala
    classification   *see* Appendix
    cystacanth   136
    eggs   106
    excretory system   86
    excystation   136
    keratin   64
    nervous system   161, 163
    osmoregulation   91
    sense organs   165
    sexual reproduction   105
Acanthocephalus lucii   149
A. ranae   149
Acanthocotyle elegans   164
A. lobianchi   164
acanthella   112
Acanthoparyphium   147
acanthor   112
Acanthorhynchus borealis   149
Acanthostomum   13
acetate transport   32
acetylcholine transport   170
acetylcholinesterase   161, 170
N-acetyl glucosamine   29
active transport   24, 25
adenine   34, 74
aerobiosis   53, 139
alanine   30, 87
Alaria   147
alimentary canal of helminths   6–17
alkaline phosphatase   86
Allocreadium lobatum   164
D-allose   27
Alosa   172
amino acids   14, 16, 92
    biosynthesis   66
    content   63, 65–67
    excretion   86
    free pool   65
    imbalance   37
    metabolism   68–69
    osmoregulatory role in monogeneans   93

transport   30–32
α-aminoisobutyric acid   37
α-amino-β-oxybutyric acid   73
aminotransferases   67
ammonia   87–89
amoeba   3, 188
Amphibdella flavolineata   164
amphids   119, 168, 170
α-amylase   40
amylopectin   43
anaerobiosis   54, 140, 139
Ancylostoma   17
    exsheathment   137
    host location   120
    osmoregulation   91
A. caninum
    crowding effect   154
    immunity   196
    skin penetration   124
    vaccination   197
A. ceylanicum   124
A. duodenale   124
A. tubaeformae   124
Angiostrongylus cantonensis   42
1,5-anhydro-D-mannitol   27
Annelida   69 (*see also* Appendix)
antibodies   177, 178
antienzymes   40–41
antigenic variation
    in malaria   181
    in trypanosomes   185–187
antigens   177
antimycin A   57
Apatemon gracilis   86
Arachnida   *see* Appendix
arginase   88
arginine   30
arsenite   28
Arthropoda   *see* Appendix
arthus   199
Ascaridia galli   42, 43
Ascaris   2
    cuticle   21

219

Ascaris—continued
  feeding 17
  haemoglobin 59
  skin test 198
  toxin 202
A. suum
  antienzymes 41
  carbohydrate transport 27
  immunity 196
A. lumbricoides
  acetylcholinesterase 170
  aminotransferases 67
  ammonia formation 88
  ascarosides 43
  carbohydrate metabolism 46, 49, 51
  collagen 64
  cytochromes 56, 57, 69
  disaccharide metabolism 47
  fatty acids 46, 47
  fermentations 49, 50
  GABA 170
  glycogen 42
  glyoxylate cycle 61
  haemoglobin 59
  immunity 196
  lipid biosynthesis 72
  locomotion 172
  malate dehydrogenase 49
  "malic enzyme" 48, 49
  metabolism 140
  migration 142
  neuromuscular control 170
  oxygen tolerance 55
  Pasteur effect 55
  pentose phosphate pathway 60
  succinoxidase system 57
  trehalose content 43
  tricarboxylic acid (TCA) cycle 53
ascariasis 199
ascarosides 43
ascarylose 43
asexual reproduction 94–97
Assiminea japonica 117
A. latericea miyazakii 117
A. parasitologica 117
Austrobilharzia terrigalensis 117
A. variglandis 117
azide 57
Azygia 13

Babesia
  immunity 183
  immunosuppression 196

  invasion of tissues 151
B. argentina 197
B. bigemina 183
B. bovis 183
B. divergens 183
B. equi 183
B. microti 183
B. rhodaini 37, 183
babesiasis 3, 199
Balantidium coli 88, 90
BCG vaccine 198
behaviour patterns 172
bile salts
  in hatching and excystation 137–139
  in lipid transport 32, 33
Biomphalaria boissyi 117
B. glabrata 117
blood feeding
  in Digenea 13
  in *Fasciola* 14–16
  in Monogenea 9
  in Schistosomes 14–16
Bolbosoma 161
Bradynema 17, 21
Brugia malayi
  circadian rhythm 127
  vaccination 197
B. pahangi
  immunity 196
  lactic acid production 45
Buglossidium luteum 113, 114
Bulinus 117

Calliobothrium verticillatum 27, 28, 90
Camallanus trispinosus 59
Capillaria obsignata 197
carbohydrates
  catabolism 45–47
  stores 42–43
  transport 26–29
carbon dioxide fixation 48–51
cardiolipin 71, 73
carotenoids 72
carriers (in transport) 73
castration 202
Catatropis 147
cell mediated immunity 177, 180
Centrovitus 147
cercaria 112, 121
  aerobiosis 140–141
  chemotaxis 118
  host location 118–119
  host penetration 122–123

cercaria—*continued*
  longevity 173
  morphology 121
  sense organs 165
Cercaria emasculens 202
C. lophocerca 202
C. milfordensis 202
Cercaria X 147, 148
Cestoda
  asexual reproduction 97
  egg hatching 134
  host location 119
  immunity 192–193
  larvae 134
  nervous system 160–161
  pathogenesis 201–202
  sense organs 165
  sexual reproduction 104–105
classification of parasites *see* Appendix
Chagas disease 30
chemotherapy 3
chemotaxis
  in cercariae 118–119
  in miracidia 115–118
  in oncomiracidia 113–115
chemotrometer 116
chloramphenicol 68
cholesterol 32, 72, 78, 122
choline 71
chymotrypsin 40
ciliates
  contractile vacuoles 83
  fatty acid production 46
  oxygen tolerance 55
circadian rhythms
  in helminths 126–128
  in Protozoa 125–126
citric acid cycle *see* tricarboxylic acid
  cycle
Clonorchis sinensis 118
Coccidia
  circadian rhythms 126
  immunity 188
coccidiosis 3
Coelenterata *see* Appendix
Coelotropha durchoni 108
coenurus 95, 97
collagen 64
Columbella lunata 117
complement 176, 199
contractile vacuoles 82–83
coracidium 105, 112, 114, 139
Crabtree effect 55

Crithidia fasciculata
  acetate synthesis 73
  amino acid biosynthesis 66
  carbohydrate metabolism 46
  ferritin transport 37
  lipid biosynthesis 72, 73
  mitochondrial protein synthesis 68
  osmoregulation 90
  ribosomes 68
C. oncopelti
  bipolar body 66
  in vitro culture 66
  methionine requirement 66
  protein synthesis 68
crowding effect 154, 155
Crustacea *see* Appendix
Ctenophora *see* Appendix
cyanide 54, 57
Cyathocotyle bushiensis 133, 138
cycloheximide 68
cycloleucine 37
Cyclophyllidea (eggs) 105, 134
cystacanth 112, 136, 139
cysteine 30
cysticercoid 97, 112, 134, 193
cysticercosis 3
cysticercus 112
Cysticercus pisiformis 138
C. tenuicollis 88, 90
cytochromes 54, 55, 56, 69
cytostome 35–36

Dactylogyrus 103
Davainea proglottina 156
delayed hypersensitivity 199
2-deoxyglucose 28, 29
6-deoxyglucose 27, 28
Derogenes 147
desaturation of fatty acids 73
Dibothriocephalus 156
Diclidophora denticulata 92
D. merlangi 9, 10, 11, 92, 145, 164
Dicrocoelium dendriticum 49, 51, 171
Dictyocaulus filaria 197
D. viviparus 49, 197
Dictyocotyle 103
D. coeliaca 92
diffusion 23
Digenea
  asexual reproduction 94–97
  eggs 103–104
  gut 11–16

Digenea—*continued*
  host location 115–119
  immunity 189–191
  nervous system 160–161, 163
  pathogenesis 201
  sclerotin 64, 65
  sense organs 164–165
  sexual reproduction 103–104
  site selection 146–147
digestion 9–17
digestive enzymes 10–11, 14–16, 39–41
2,4-dinitrophenol 27
Dipetalonema viteae 44
Diphyllobothrium 136
D. dendriticum 108
D. latum
  acetylcholinesterase 170
  anaemia 202
  excretory system 84
  vitamin B_{12} 35, 202
Diplodiscus 147
D. subclavatus 13
Diplostomum gasterostei 133, 147
D. phoxini 133, 142, 147, 170
D. spathaceum 133, 142, 147–148, 164
Diplectanum aequans 145
Diplozoon paradoxum 145
Dipylidium caninum 170
Dirofilaria immitis
  annual cycle 128
  carbohydrate metabolism 51
  circadian rhythm 127
  immunity 196
  vaccination 197
disaccharides 43
Discocotyle sagittata 92, 145
DNA 74–81, 141
drug resistance 3

Echinococcus granulosus
  acetylcholinesterase 170
  asexual reproduction 95, 97
  bile salts 138
  development 157
  ethanol production 47
  excretory system 84
  excystation 134
  growth of hydatid cyst 156–158
  hexokinases 47
  immunity 192
  migration 142
  neurosecretion 171
  pentose phosphate pathway 60

  site selection 149
  skin test 198
  urea and uric acid 88
  vaccination 198
E. multilocularis 198
Echinoparyphium serratum 138
Echinorhynchus truttae 136
Echinostoma 147
E. revolutum 86
eggs
  Acanthocephala 106
  Cyclophyllidea 105, 134–135
  Digenea 103–4
  Monogenea 103
  Nematoda 107
  Pseudophyllidea 105
egg production in schistosomes 14
Ehrlich ascites tumour cells 25
eicosadienoic acid 73
eicosenoic acid 73
Eimeria
  bile salts 138
  excystation 132
  hyperplasia 202
  immunity 188
  invasion of tissues 151–153
  oocyst 135
  triglycerides 71
E. bovis 188
E. meleagrimitis 188
E. tenella 132, 188
electron transport 55–58
endocytosis 11, 35–37
Enoplus brevis 60
Entamoeba histolytica
  amino acid biosynthesis 66
  cholesterol 72
  excystation 132
  immunity 188
  lysosome 37
Enterobius vermicularis 126
Entobdella hippoglossi 9, 92
E. soleae
  alimentary canal 9
  amino acid content 92
  host location 113–115
  sense organs 162
Entodinium caudatum 69, 90
E. shaudini 67
entomophilic nematodes 17, 21
eosinophils 194, 195, 199
ergosterol 72, 78
esterases 13

ethanolamine 71, 73
Eurytrema 147
Eustrongyloides ignotus 55
excretory systems 82–87
excystation 130–137
exsheathment 137

facilitated diffusion (mediated transport) 23–25
Fasciola hepatica
  acetylcholinesterase 170
  amino acid transport 31
  aminotransferases 67
  ammonia production 88
  anaemia 201
  bile salts 138
  carbohydrate metabolism 46, 49
    transport 26, 28–29
  cyst wall 133
  cytochromes 56
  egg shell 104
  excretory system 84–86
  fascioliasis 3
  fatty acids 33–34, 46
  feeding 12–16
  glycogen 43
  glyoxylate cycle 61
  gut 13
  hatching 131
  hyperplasia 202
  immunity 189
  lipid synthesis 73
  metabolism 140
  metacercaria 131
  osmoregulation 90
  penetration of host 122
  phospholipid synthesis 73
  pinocytosis 35
  sense organs 164
  surface (tegument) 18–19
  TCA cycle 53
  transmission 115–116
F. gigantica 88, 90
Fascioloides magna 117, 122
fatty acids 32–34, 46–47, 70–71, 119, 122
fermentation 49–51
filarial worms 107, 125–128, 198
filariasis 3
flagellates 2, 83
flame cells 83–84
flavoproteins 55–58
flavoprotein oxidases 54
Fick's law 23

fructose 27, 29, 40, 43, 47
fucose 27

galactose 28, 29, 39, 43
Gasterostomata 11–12
Gastrocotyle trachuri 172
gastrodermis
  in Digenea 13–16
  in Nematoda 17
Gigantobilharzia huronensis 117, 118
glucosamine 29
glucose 27–29, 43
glutamate 30
glyceraldehyde-3-phosphate dehydro-
    genase 44, 45
glycerides 71
glycerol 28, 29, 71, 73
α-glycerophosphate 28, 58
glycine 30, 92, 93
glycocalyx 19–20
glycogen 42–43, 44
glycolysis 43–44
glycosides 43
glyoxylate cycle 61
Glypthelmins quieta 115, 147
Gorgodera amplicava 118
Gorgoderina vitelliloba 13
gregarines 37, 94, 99
guanine 34–35, 74
Gyrodactylus 103, 164

haematin 10, 14
haematin cell 10–11
Haematoloechus medioplexus 13, 35
haemoflagellates 37
haemoglobin 11, 14, 15, 16, 35, 58–60, 64, 69
Haemonchus contortus
  anaemia 201
  annual cycles 128
  carbohydrate metabolism 49
  exsheathment 137
  haemoglobin 59
  immunity 194
  larval behaviour 173
  metabolism 140
  neurosecretion 172
  self cure 194
  sense organs 168
  vaccination 197
Haemoproteus 108
Haplometra cylindracea 13
H. intestinalis 115

hatching mechanisms   130–137
Halipegus   147
Hammaniella   161
Helisoma anceps   117
H. trivolvis   117
Hemiurus   147
Heronimus chelydrae   164
Heterakis   132
Heterodera   107, 173
Hexabothriidae   9
hexacanth   105
hexokinases   47
Himasthla quissetensis   84, 141
H. secunda   202
histamine   195, 198
Histomonas meleagridis   132
Holostephanus luhei   86, 133, 138
hookworms   3, 17
   host location   120
   larva migrans   142
   migration   144
   skin penetration   124
   skin reactions   198
host-parasite interface   5
host specificity   114–120
humoral immunity   177, 180
hydatid cyst   95, 97, 112
hydrogen peroxide   54, 57
5-hydroxytryptamine (5-HT)   170
Hymenolepis citelli
   amino acid transport   32
   aminotransferases   67
   destrobilation   156
H. diminuta
   acetylcholinesterase   170
   amino acid content   92
   amino acid fluxes   36–38
   amino acid transport   31–32, 39
   aminotransferases   67
   ammonia content
   ammonia production   88
   α-amylase (contact digestion)   40
   antienzymes   41
   bile salts   138
   carbohydrate metabolism   46
   carbohydrate transport   27–29, 39
   carbon dioxide fixation   49
   circadian migration   143–144
   crowding effect   154, 155
   cysticercoid   135
   cytochromes   56
   egg structure   135
   excretion   86

   fermentations   49–51
   glycogen utilisation   43
   growth   156
   immunity   92–93
   lipid biosynthesis   72–73
   lipid transport   32–33
   "malic enzyme"   49
   neurosecretion   170
   osmoregulation   90
   permeability to fructose   47
   phospholipid biosynthesis   73
   phosphohydrolases   40
   pinocytosis   35
   protonephridia   86
   purine transport   33–34, 39
   pyrimidine transport   33–34, 39
   ribonucleases   40
   site selection   143–144, 149, 154
   succinate excretion   49
   TCA cycle   53
   triglyceride synthesis   73
   vitamin transport   35
H. fausti   148
H. microstoma
   α-amylase (contact digestion)   40
   carbohydrate transport   27–28
   crowding effect   154
   immunity   192
   site selection   148
H. nana
   adrenaline   170
   aminotransferases   67
   growth   156
   immunity   192
   migration   144
hyperplasia   201–202
hypoxanthine   33, 34, 35

Ichthyophthirius multifiliis   90
immediate hypersensitivity   198
immunisation   197–198
immunity   175–196
immunodiagnosis   182
immunoglobulins   177–180
   IgA   177, 179, 193, 195, 196
   IgD   177, 179
   IgE   177, 179, 189, 193, 194–196, 198
   IgG   177, 179, 185, 189, 193, 195, 196, 198
   IgM   177, 179, 189, 195, 196
immunopathology   198–200
Insecta   *see* Appendix
invasion processes   150–153

in vitro culture 3, 6
  essential amino acids 67
  Fasciola hepatica 15
  Hymenolepis diminuta 67
  Schistosoma mansoni 14
iodoacetate 27
isocitrate lyase 61
isoenzymes 47
isoleucine 30
Isospora 126, 132
I. arctopitheci 132
I. bigemina 132
I. canis 132
I. endocallimici 132

keratin 64
α-ketoglutarate 67
kinetoplast 75, 76, 77
Krebs Cycle (*see* tricarboxylic acid (TCA) cycle)

Lacistorhynchus tenuis 87, 90, 91, 156
lactic acid 45–46, 86
larva migrans 142
Lasiotocus 147
Leishmania 187–188
L. braziliensis 187–188, 198
L. donovani
  amino acid biosynthesis 66
  aminotransferases 67
  arginine biosynthesis 66
  immunity 187–188
  lipid biosynthesis 72
  vaccination 198
L. mexicana 187–188
L. tarentolae
  amino acids 67
  arginine synthesis 66
L. tropica 187–188, 198
leishmaniasis 199
Leptococtyle minor 164
Leuchochloridium paradoxum 202
leucine 30
Leucocytozoon 108
Ligula intestinalis 136, 202
Limanda aspersa 172
L. limanda 113, 114
linoleic acid 73
lipase 40
lipid
  biosynthesis 90–91
  catabolism 73–75
  classification 70

  parasite composition 70–71
  transport 32–34
Litomosoides carinii 55, 173, 196, 197
Littorina littorea 202
Loa loa 127
Lota 149
Loxogenes 147
lymphocytes 177, 180
lymphokine 182
lysine 30
lysolecithin 73

Macracanthorhynchus 161
Macracanthorhynchus hirudinaceus 32, 106
malaria (also *Plasmodium*) 2, 3, 4
  asexual reproduction 96, 98
  circadian rhythms 125–126
  free pool of amino acids 66
  glomerulonephritis 199
  haemolytic anaemia 199
  immunity 181–183
  malarial pigment 70
  red cell invasion 152
  sexual reproduction 98
  TCA cycle 53
  vaccination 198
malate dehydrogenase 48
malate synthetase 61
"malic enzyme" 48–50
mannose 29, 43
Mazocraes alosa 172
Megalodiscus temperatus 135
Meloidogyne 107
p-mercuribenzoate 27
Mermis nigrescens 17
M. subnigrescens 69
Mesocoelium 147
Mesozoa *see* Appendix
metacercaria 112
  cyst wall structure 133
  encapsulation 176
  hatching 133
methionine 30, 39, 66
α-methyl glucose 27
3-O-methyl glucose 27, 29
micelles 32
Michaelis-Menten enzyme kinetics 24, 25–26, 31
Microchirus (Solea) variegatum 114
Microcotyle 8
Microcotylidea 9
microfilariae 127
Microphallus pygmaeus 88

microtriches 18, 19
migration 141–145
miracidium 112
  aerobiosis 140, 139
  chemotaxiis 115–118
  host location 115–118
  host penetration 120–122
  longevity 173
  morphology 121
  sense organs 164, 167
Mollusca *see* Appendix, 53, 115–117, 202
Moniezia benedeni 87
M. expansa 2
  α-amylase 40
  carbohydrate metabolism 49
  cytochromes 56, 69
  excretory system 84–85
  osmoregulation 90
  TCA cycle 53
Moniliformis dubius
  amino acid transport 32
  α-amylase 40
  bile salts 138
  carbohydrate metabolism 46, 49, 51
  carbohydrate transport 27, 29
  cytochromes 56
  disaccharide metabolism 47
  ethanol production 46
  excystation 136
  fatty acid production 46
  interspecific interactions 154
  "malic enzyme" 49
  sexual dimorphism in glycogen
    stores 42
  site selection 149
  urea cycle 89
Monocotylidae 9
Monogenea
  alimentary canal 6–9
  eggs 120
  host location 114–115
  nervous systems 160, 163
  sense organs 162–164, 166
  sexual reproduction 100–103
  site selection 145–146
Monopisthocotylea 9
monosaccharides 27–30, 39, 43
Mustela nivalis 154
Mytilus edulis 202

Necator 120
N. americanus 124, 141

Nematoda
  acetylcholinesterase 170
  alimentary canal 16–17
  amino acid transport 32
  classification *see* Appendix
  crowding effect 154
  cuticle 21
  eggs 107, 137
  excretory system 87
  exsheathment 137
  haemoglobin 58–59
  host location 119–120
  immunity 193–196
  keratin 64
  larvae 112
  larval behaviour 173
  moulting 158, 172
  muscle cell 171
  nervous systems 161, 163
  osmoregulation 91
  penetration mechanisms 124–125
  sense organs 168–169, 165–170
  sexual reproduction 102, 106–107
  TCA cycle 53
Nematomorpha *see* Appendix
Nematospiroides dubius 144
Nemertea *see* Appendix
Neoechinorhynchus 149, 154
neoplasia 201–202
nephrostome 84
Nereis diversicolor 108
nervous systems 160–163
neurosecretion 171–172
nicotinamide 35
Nippostrongylus brasiliensis
  carbohydrate metabolism 49, 51
  cuticle 21
  fatty acid composition 70
  haemoglobin 59
  host location 120
  immunity 194–195
  metabolism 141
  osmoregulation 91
  Pasteur Effect 55
  self cure 194
non-productive binding (in transport) 29, 34
Notocotylus 147
N. attenuatus 133
nucleic acids 75–80
Nudacotyle 147
Nyctotherus 108

oesophageal glands 8–11

Onchocerca cervicalis 128
O. gutterosa 128
onchocerciasis 3
Oncomelania nosophora 117
oncomiracidium 112, 113–115, 172
oncosphere 112
Oochoristica symmetrica 156
Opalina 108
Opisthoglyphe ranae 13
opsonins 176, 182
ornithine cycle *see* urea cycle
ornithine transcarbamylase 88
osmotrophy 35
Otodistomum 147
oxidative phosphorylation 54
oxygen 53–55, 58–60

palmitate 32–33, 73
Paragonimus 144, 147
P. ohirai 116, 117
P. westermani 202
Parascaris equorum 2, 42, 43
Paramphistomum explanatum 88
Parictotaenia paradoxa 97
Parorchis 135
P. acanthus 133
Parvatrema timondavidi 133
Pasteur Effect 55, 140
pathogenesis 198–202
Pelodera 173
Pentastomida *see* Appendix
pentose phosphate pathway 60–61
perch 149
permeases 23
phagocytosis 153, 176
phagotrophy 35
pharyngeal glands 9
phasmids 119, 168, 170
phenylalanine 30
Philophthalmus megalurus 164
phosphatidylcholine 71
phosphatidylethanolamine 71
phosphatidylglycerol 71
phosphatidylinositol 71
phosphatidylserine 71
phosphoenolpyruvate carboxykinase
    (carboxylase) 48–51
phosphohydrolases 40
phospholipids 70, 71, 73
photoreceptors 161–170
Phyllobothrium 149
Physa gyrina 115, 117
pinocytosis 35–37

pinworms 3, 126, 149–150
Plagiorchis 147
plant-parasitic nematodes 3, 107, 119, 173
plasmalogen 71
plasma membrane 18–19
Plasmodium (*see also* malaria)
    cholesterol 72
    immunity 181–183
    immunosuppression 183
    invasion of red cells 150–152
    lipid biosynthesis 72
    pinocytosis 37
    reproductive synchrony 108
    toxin 183
P. berghei
    ammonia production 88
    carbohydrate metabolism 46
    exflagellation 125
    immunity 181–182
    pentose phosphate pathway 60
    protein synthesis 68
P. knowlesi
    amino acid biosynthesis 68
    invasion of red cells 151
    protein synthesis 68
    schizogony 125
P. cathemerium 125
P. chabaudi 125
P. cynomolgi 125
P. falciparum 183
P. gallinaceum 69
P. lophurae
    amino acid metabolism 69
    ammonia production 88
    invasion of red cells 151
P. malariae 125
P. vinckei 181
Platyhelminthes, classification *see*
    Appendix
plerocercoid 112, 139
Pleurogenes 13
Pleuronectes platessa 113, 114
Podocotyle 147
polycercus 97
Polymorphus minutus
    bile salts 138
    carbohydrate transport 27, 29
    carotenoids 72
    cystacanth 135
    egg production 106
    excystation 136
    nervous system 161
    site selection 149

Polyopisthocotylea 9, *see also* Appendix
polysaccharide 42–43
Polystoma integerrimum
  alimentary canal 8
  life-cycle 145–146
  reproductive synchrony 108, 145–146
  site selection 145–146
Pomphorhynchus 149
Porifera *see* Appendix
Postharmostomum 147
Posthodiplostomum 133, 144
P. cuticola 118, 147
procercoid 112, 139
proline 30, 39, 92, 93
1,2-propanediol 28
Prosostomata 12
Protancyrocephalus 172
protein
  biosynthesis 68
  composition 63–65
  determination 64
  metabolism 68–69
Proteocephalus filicollis 149, 154
protonephridia 83–86
Protozoa
  asexual reproduction 94–95
  classification *see* Appendix
  excretion 83, 88
  immunity 181–189
  nutrition 26–30, 35
  osmoregulation 82–83, 89–90
  sexual reproduction 98–100
Pseudanthobothrium 149
Pseudophyllidea
  eggs 105
  plerocercoids 131, 136, 139, 202
  sclerotin 65
Psilotrema oligoon 133
purines 34–35, 74–75
pyridoxine 35
pyrimidines 34–35, 74–75
pyruvate kinase 51
pyruvic acid 46, 49

quinone tanning 64–65, 103, 104

Raia radiata 149
Raillietina cesticillus
  glycogen utilisation 42
  growth 156
  immunity 192
  site selection 148
Rana clamitans 127

reaginic antibody (reagin) 189, 194
redia 96–97, 112, 139
reproductive synchrony 107–108
respiratory pigments 58–60, 69–70
reticuloendothelial system 176
ribose 29, 43, 74
riboflavin 35
ribonucleases 40
Rotifera *see* Appendix

Sacculina parasitica 202
Sanguinicolidae 13, 147
Sarcocystis 201
scabies 3
Schistocephalus solidus
  growth 156
  metabolism 141
  osmoregulation 90
  plerocercoid 136
Schistorchis carneus 13
Schistosoma bovis 126
S. haematobium
  cercarial emergence 126
  circadian rhythm 126
  granuloma formation 191
  hyperplasia 202
  immunity 191
  neoplasia 202
S. japonicum
  aminotransferases 67
  cercarial emergence 126
  granuloma formation 191
  host location 117
  immunity 191
S. mansoni
  acetylcholinesterase 170
  amino acid content 92
  amino acid transport 31, 39
  ammonia formation 88
  anaemia 201
  carbohydrate metabolism 49, 51
  carbohydrate transport 26–27, 28, 29, 39
  cercarial emergence 126
  chemotaxis 116–118
  feeding 13, 14–15
  fermentation 50
  granuloma formation 191
  hexokinases 47
  host location 116, 117, 118
  immunity 189–191
  lactic acid formation 45–46
  lipid biosynthesis 72–73

S. mansoni—continued
  metabolism   140
  migration   144
  Pasteur effect   55
  penetration of host   120–123
  pinocytosis   35
  proteolytic enzymes   21–22
  purine transport   39
  pyrimidine transport   39
  sense organs   164
  surface structure   20
  TCA cycle   53
  tegument   20
  transformation of cercariae   123
S. mattheei   197
S. rhodaini   126
Schistosomattium douthitti   117, 118, 119
schistosomiasis   3, 4, 15, 191, 199, 200
schistosomes
  catecholamines   170
  circadian rhythms   126
  dermatitis   143
  5-hydroxytryptamine   170
  immunity   189–191
  self cure   190
  site selection   147
  skin reactions   198
  toxin   201
  transformation of cercariae   123
schizogony   96, 98, 99
scleroproteins   64–65
sclerotin   64–65
self cure   190, 194–195
sense organs   161–170
serum sickness   199
sexual reproduction
  Acanthocephala   102, 105–106
  Cestoda   101, 104–105
  Digenea   101, 103–104
  Monogenea   100–103
  Nematoda   102, 106–107
  Protozoa   98–100
Silurus   149
site selection   145–150
Skrjabingylus nasicola   154
sodium ion requirement   29, 32
Solea solea   113, 114, 115
sphingolipid   71
Spirometra mansonoides   202
Spirorchis   164
sporocyst
  Digenea   96–97, 112, 139–140
  Protozoa   132

Stephanostomum   147
sterols   72, 78
Stieda body   132
Strigeida   131, 133
strobilocercus   97
Strongyloides fulleborni   124
S. ratti   71, 141
Succinea putris   202
succinic acid production   45–47, 51
succinoxidase system   57
surface enzymes   39–40
surface morphology   18, 19–21
Syngamus trachea   59
Syphacia muris   49, 126

Tachygonetria   149–150
Taenia crassiceps   27, 28, 32
T. hydatigena   56, 192
T. multiceps   97, 158
T. ovis   192, 197
T. pisiformis   134, 192
T. saginata   134, 192
T. serialis   158
T. taeniaeformis
  acetylcholinesterase   170
  ethanol production   47
  neoplasia   202
  urea   88
  uric acid   88
taeniasis   3
tegument, of platyhelminths   17–22
tetraethylthiouram   73
Tetraonchidae   9
Tetraonchus monenteron   7, 145
theileriasis   3
thiamine transport   35
thymine transport   34, 74
threonine   73
tick-borne diseases   3
toxins   183, 201
Toxocara cati   43
toxocariasis   199
Toxoplasma gondii   132, 151–153, 188
toxoplasmosis   199
Trachurus trachurus   172
transcuticular absorption in nematodes
    see entomophilic nematodes
Transversotrema patialense   89
trehalose   43
tricarboxylic acid cycle (TCA)   52–53
Trichinella spiralis
  carbohydrate metabolism   49
  eggs   107

Trichinella spiralis—continued
  fatty acid production   47
  immunity   195–196
  skin test   198
Trichobilharzia elvae   117, 118
T. physellae   117
trichomonads   55, 57, 72
Trichomonas foetus   198
T. gallinae   53
T. vaginalis   68–69
trichomoniasis   199
Trichostrongylus   107
T. axei   137
T. colubriformis   197
Trichuris   17
T. muris   196
T. vulpis   27
Trypanosoma brucei   46, 58, 66
T. congolense   57, 186
T. cruzi
  amino acid biosynthesis   66
    metabolism   68
    transport   30
  aminotransferases   67
  cytochromes   57
  ergosterol   72
  immediate hypersensitivity   198
  pentose phosphate pathway   60
  vaccination   198
T. equinum   185–186
T. equiperdum   26, 29
T. gambiense
  alanine biosynthesis   87
  amino acid biosynthesis   66
    transport   30
  carbohydrate metabolism   46
    transport   26, 29
  immunity   185
  osmoregulation   90
T. lewisi
  carbohydrate metabolism   46
    transport   26, 29
  ergosterol   72
  growth *in vivo*   186
  immunity   184–185
T. musculi   185

T. rhodesiense
  amino acid biosynthesis   66
    metabolism   69
  cytochromes   57
  ergosterol   72
  immunity   185
T. rotatorium   127
T. vivax   57, 186
trypanosomes
  acetate   46, 73
  antigenic variation   186
  asexual reproduction   94
  circadian rhythm   127
  cytochromes   57
  free pool of amino acids   66
  glyceride content   71
  α-glycerophosphate oxidase   58
  hexokinases   47
  immunity   183–187
  kinetoplast DNA   78–79
  lipid biosynthesis   72
  TCA cycle   53
trypanosomiasis   3, 199
trypsin   41, 132, 133, 134, 137
tryptophan   30, 39
tyrosine   30

uracil   34, 74
urea   83, 87–89, 91
urocystidium   97
urocystis   97

Vahlkampfia calkensis   90
valine   30
vitamins   35, 202

waxes   72
World Health Organization (WHO)   3
Wuchereria   17
W. bancrofti   127

Xiphidiocercariae   133
xylose   29, 43

Zoogonus rubellus   141
Zoogonoides laevis   147, 117